Nietzsche and Embodiment

SUNY series in Contemporary Continental Philosophy
Dennis J. Schmidt, editor

Nietzsche and Embodiment

Discerning Bodies and Non-dualism

Kristen Brown

State University of New York Press

Published by
State University of New York Press, Albany

© 2006 State University of New York

All rights reserved

Printed in the United States of America

No part of this book may be used or reproduced in any manner whatsoever without written permission. No part of this book may be stored in a retrieval system or transmitted in any form or by any means including electronic, electrostatic, magnetic tape, mechanical, photocopying, recording, or otherwise without the prior permission in writing of the publisher.

For information, address State University of New York Press,
194 Washington Avenue, Suite 305, Albany, NY 12210-2384

Production by Diane Ganeles
Marketing by Susan M. Petrie

Library of Congress Cataloging-in-Publication Data

Brown, Kristen, 1964–
Nietzsche and embodiment : Discerning bodies and non-dualism / Kristen Brown.
p. cm. – (SUNY series in contemporary continental philosophy)
Includes bibliographical references (p.) and index.
ISBN-13: 978-0-7914-6651-3 (hardcover : alk. paper)
ISBN-10: 0-7914-6651-5 (hardcover : alk. paper)
ISBN-13: 978-0-7914-6652-0 (pbk : alk. paper)
ISBN-10: 0-7914-6652-3 (pbk : alk. paper)
1. Nietzsche, Friedrich Wilheim, 1844–1900. 2. Body, Human. 3. Human beings. 4. Self. 5. Women—Physiology. I. Title. II. Series.
B3317.B73 2006
193—dc22

2005006557

10 9 8 7 6 5 4 3 2 1

For my family

Contents

Acknowledgments

A project such as this comes about through the support of many people along the way. There are a few to whom I would like to express special thanks: to Michael Galchinsky, Anne MacMaster, Lisa Sigel, and Steve Smith for their helpful comments on certain draft chapters; to David Garcia, John Thatamanil, and Nona Olivia for their friendship, and comments on versions of the manuscript; and to Nona and her sons, Jasper Segal and Scott Vickers, for opening their home to me; a similar thanks to Idit Dobbs-Weinstein and Sophia Dobbs. Finally, thank you to the Brown, Garcia, and Golden families, and especially Bruce and Chelsea, whose nurturance helped make this possible.

An earlier version of chapter 2 appeared as "Possible and Questionable: Opening Nietzsche's Genealogy to Feminine Body" in *Hypatia* volume 14 number 3, 1999.

Abbreviations

Translations of Nietzsche's works are abbreviated as follows:

A	*The Anti-Christ*, in *Twilight of the Idols/The Anti-Christ*, translated by R. J. Hollingdale (London & New York: Penguin Books, 1968).
BGE	*Beyond Good and Evil*, translated by Walter Kaufmann (New York: Random House, 1966).
BT	*The Birth of Tragedy*, translated by Walter Kaufmann (New York: Random House, 1967).
D	*Daybreak: Thoughts on the Prejudices of Morality*, translated by Walter Kaufmann (Cambridge: Cambridge University Press, 1982).
GM	*On the Genealogy of Morals*, translated by Walter Kaufmann and R. J. Hollingdale (New York: Random House, 1989).
GS	*The Gay Science*, translated by Walter Kaufmann (New York: Random House, 1974).
MW	On Music and Words, translated by Oscar Levy, in *The Complete Works of Friedrich Nietzsche* (New York: Russell & Russell, 1964) 2: 29–47.
PTG	*Philosophy in the Tragic Age of the Greeks*, translated by Marianne Cowan (Chicago: Henry Company, 1962–69).
TI	*Twilight of the Idols*, in *Twilight of the Idols/The Antichrist*, translated by R. J. Hollingdale (London & New York: Penguin Books, 1968).
TL	On Truth and Lying in a Non-Moral Sense, in *The Birth of Tragedy and Other Writings*, translated by Ronald Speirs and edited by Raymond Geuss and Ronald Speirs (Cambridge: Cambridge University Press, 1999).
UDH	On the Uses and Disadvantages of History for Life, in *Untimely Meditations*, translated by R. J. Hollingdale (Cambridge: Cambridge University Press, 1983).

WP *The Will to Power*, translated by Walter Kaufmann and R. J. Hollingdale (New York: Random House, 1968).

Z *Thus Spoke Zarathustra*, translated by Walter Kaufmann (New York: Penguin Books, 1978).

References to English translations of Nietzsche's writings appear in parentheses with abbreviated titles followed by aphorism number. When necessary, the aphorism number will be preceded by the chapter number or abbreviation of chapter title. For those texts not written in aphorism format, page numbers will be given.

When the translation is mine I refer to Nietzsche's works parenthetically with "KSA" followed by a volume number and page number referring to *Friedrich Nietzsche: Sämtliche Werke, Kritische Studienausgabe in 15 Einzelbänden*, edited by Giorgio Colli and Mazzino Montinari (Berlin: de Gruyter, 1988).

Translations for Merleau-Ponty's works are abbreviated as follows:

N *Nature: Course Notes from the Collège de France*, compiled and with notes by Dominique Séglard. Translated by Robert Vallier (Evanston, Ill.: Northwestern University Press, 2003).

PP *Phenomenology of Perception*, translated by Colin Smith (London: Routledge & Kegan Paul, 1962).

CHAPTER 1

Introduction: Nietzsche and Embodiment

PART I. OVERVIEW

We often find ourselves theorizing about why we feel a certain way. We try to guess about the sources of an ailment that we or someone close to us may have. In the case of a developing headache, we might ask, "Is the onset of this pain the result of dehydration perhaps? Or mental stress? Is it the result of environmental or psychological circumstances? Or maybe it's from a blend of these?" We might test the first theory and drink a big glass of water, or we might explore the second by taking a long walk or resting to try to calm our mind. Just how we make sense of the origin of a headache or origins of other sorts of human experiences—not just ailments or feelings but concepts and beliefs too—is a topic that occupies the attention of Friedrich Nietzsche in many of his writings. Nietzsche, one could say, has a story to tell about origins and especially about those origins that might help him to make sense of the often confusing and complex relations we inhabit across felt and imagined human experiences, that is, across what have come to be called respectively bodily and mental experience.

Nietzsche's medium for investigating origins of our felt and imagined experience is written language. According to psychologist Lisa Capps and linguist Elinor Ochs, "language is the greatest human resource for representing and structuring events in our lives. And no language practice has more impact in this direction than storytelling" (Capps & Ochs 1995, 13). It might seem strange that a text about the writings of Nietzsche and the idea of embodiment should begin by discussing language practice and storytelling. And yet, many commentators on Nietz-sche's writings have remarked that for Nietzsche, the activity of writing and storytelling not only creates literature but also enacts and creates a life—indeed, an embodied life.[1]

Nietzsche's writings indicate his relentless resistance to traditional and received categories of being. This resistance extends to the spheres of felt and imagined experience—traditionally distinguished in the West as "body" and "mind." With respect to these spheres, language has generally been perceived as residing in the province of "mind." If in fact, one questions the often-accepted borders of mind and body, as does Nietzsche, and entertains a variety of arrangements configuring new and often highly integrated relations between so-called "mind" and "body," then attention to the concept of language in a book about embodiment seems appropriate. This is because such attention may simultaneously enact attention to the concept of body. In this context, one might think of the position of Maurice Merleau-Ponty for whom language in *Phenomenology of Perception* (1945/1962) is one of many behavioral processes integrated together and forming a whole. The whole constitutes an individual's experience, which Merleau-Ponty calls "phenomenological body."

A potent form of language is storytelling. Storytelling is said to be universal among all human cultures (Capps & Ochs 1995, 13). It is a potent means of making sense of, investigating, and creating human experiences and societal norms. Our stories, contend Capps and Ochs, are not diminished versions of "what actually happened," but in an important way, indicate and shape who we are and are becoming. Our stories recruit our evolving images and recollections of an event. They make and present us as they make and present our individual and collective theories of reality. "It is not what 'really happened' but rather experiencers' theories of what happened that provide continuity between past, present, future and imagined lives" (Capps & Ochs 1995, 21).

Stories bear different images or memories of an event. Some may be laden with worn images plucked again and again by the narrator(s). These repeated renditions of a story resemble and co-assemble what may come to be considered the "official story" of the event. Some stories, on the other hand, expose forgotten or obscured impressions not typically associated by a narrator or community with the event. Such stories—distinguished by their fresh configurations—unearth and pose what can be called with respect to the event, a "subjugated story."

In the context of nineteenth-century Europe, Nietzsche's story about origins of human felt and imagined experience could be said to be a subjugated story. It uncovers generally buried threads. As Nietzsche finds these threads, he usually does not weave them together into a neat pattern and leave them there intact. Instead he weaves a position and submits it to stress. Often the pattern and images come undone, leaving some threads available to be taken-up and repositioned by another of Nietzsche's endeavors. The new pattern generally differs from the previous one, even if its very possibility requires for its construction the pliable elements and fertile context of the recently twisted and abandoned position.

Nietzsche's story of origins dusts off and generates often-ignored images and overlooked combinations of images of origin-events and shows them as generative relations to one other.

Among the tales he tells, Nietzsche seems to offer none as his official story. He weaves his elaborations, entertaining one position for a while only soon to find for himself a new posture through which he might concentrate abrasive pressure on the very organizing figure or structure he had been carefully working together. Under the stress, the tale's structure or figures may fray. Its nodes and pattern of images may partially disintegrate. Nietzsche's tale telling about origins then, is also an exercise in investigative entertainment, disassembly, and creativity.

Stories can be helpful, destructive, or neutral with respect to their relative ability to make sense of the experience of a narrator or of that of members of a narrator's community.[2] For some of us living in the twentieth or twenty-first centuries, Nietzsche's story has helped or continues to help us "makes sense" of our experience of origins. One such person for whom this is the case is Michel Foucault. In Foucault's essay, "Nietzsche, Genealogy, Morality," for instance, we can see Nietzsche's story about origins helping Foucault to tell his story about origins. He leans on Nietzsche's method of excavating overlooked images or assembling atypical associations of images—a method that has come to be known as Nietzsche's "genealogical method." Via this leaning, Foucault's story with the help of Nietzsche's is able to distinguish the concept of genealogical origins from a traditional concept of Origin. Whereas the traditional concept Origin normally includes in its definition an ultimate foundation or beginning, the genealogical concept origin does not.

For others such as Robert Jay Lifton, author of *The Protean Self: Human Resilience in an Age of Fragmentation* (1993), and twentieth-century Japanese Buddhist philosopher, Shin'ichi Hisamatsu, author of *The Fullness of Nothingness* (1984) Nietzsche's story—although not directly inspiring their stories about the concept of self—lines up with them. Lifton, like Nietzsche, views the nature of human selves as fundamentally fluid and Proteus is the Greek sea god that symbolizes this. Proteus is a god of many forms—no single form suits or sustains him and it's this way for the human self too, says Lifton. According to Lifton's idea of humans as protean, we are multifaceted selves—always shifting and changing. Hisamatsu likewise sees the human concept of self exceeding any single form while ex-pressing itself in countless specific forms. In a dialogue between Hisamatsu and Paul Tillich about the relation between finite form and ultimate reality (Hisamatsu prefers to call ultimate reality: "Formless Self")—"Hisamatsu replies that in its self-concretization Formless Self can assume innumerable forms. In the same conversation, Tillich persists in thinking that the Formless Self is somehow separate from the specific forms in which it manifests itself. But the point is that the Formless Self does not have any form apart from the specific forms in which it manifests itself. . . . A specific form does not stand for or represent the ultimate; the ultimate is ex-pressed, literally pressed out or injected into, the specific form" (Stambaugh 1999, 65–66).

In light of fellow story-tellers such as Foucault, Lifton, and Hisamatsu, Nietzsche's story of origins seems to be a little less in our day, the submerged story it may have been in his. Foucault, Lifton, and Hisamatsu are just three tale-tellers among many whose stories—spanning disciplines—also show the influences of or a theoretical kinship with Nietzsche's story. In recent years many narrators across disciplinary divides have been revising received enlightenment stories about human origins. Many have become aware of certain "official storytellers" *as* "official storytellers": Kant, Rousseau, and Hegel, for instance, and, have recognized certain versions of their tales passed along by interpreters as the "official versions" but, importantly, not the only versions.

Nietzsche's story, if still a subjugated one, seems less so in the company of stories about selves and origins told by his more recent comrades; Nietzsche's way of "making sense" makes sense to at least some of our contemporaries. Although Nietzsche's story seems for some to increasingly make sense—certain lineages of philosophers, psychologists, Buddhist teachers, literary theorists, and anthropologists, for instance—"making sense" is not the same as "common-sense." As anthropologist Clifford Geertz sees it, what passes for common sense are customs agreed upon by people with shared habits, histories, and geography. Such customs are generated by particular group circumstances. Geertz insists, although "common sense" exists, universal or objective common sense does not (Geertz 1983, 75). In other words, what passes for common sense is not something that people can objectively discern across cultures independent of societal practices.

A tale-telling method—for instance, the way a narrator attempts to "make sense" of complex human experiences—may coincide with a community's view of common sense but it need not. Making sense and common sense do not necessarily share an agreed-upon view of "reality." Those of us living in industrialized communities may try to make sense of often complicated or confusing human experiences by telling certain kinds of stories. These stories often carry the mark of implied techno-societal assumptions. For instance, they may have logical structures that assume certain premises about individual responsibility and freedom, productivity, and scientific causal laws. Stories housing such assumptions generally reduce to an ultimate foundation or Origin. The ways that we in technological societies often try to comprehend our lives—that is, through appeals to idealism, Cartesian dualism and Newtonian cause-and-effect logic—can be said to overlap with what has come to be in industrialized communities, one form of common sense. This particular sort of common sense, and tales drawn to its hue, harbors in it an implicit arbiter of "reality." Saturated in dualist assumptions of Plato-Cartesian ancestry, this perceived arbiter of reality is generally associated with the European enlightenment and is sometimes called "technological reason." Nietzsche's story, although at times using certain technological images and themes—generally deflects a deeply entrenched technological view of common sense and reality coloring many of our experiences. Because of this deflection, Nietzsche's story remains promising territory for investigating ways human experience may disconfirm or exceed technological reason. *Nietzsche and Embodiment* is in fact a meditation on the possibility of discerning human experience as dynamically non-dual and Nietzsche's published writings provide focused resources for this investigation.

Just as Nietzsche has a story to tell about origins of certain embodied and reflected worlds, I have a story to tell about Nietzsche's writings. My story started out like many stories—not quite sure where it would end up. Such a start is in itself telling. My story of Nietzsche's ends with a transition to the phenomenological writings of Merleau-Ponty. One of the points I emphasize in the transition is that for Merleau-Ponty the act of speaking or writing is simultaneously the way of accomplishing the meaning of a thought. It is not a translation of a meaning that one has "already thought of," but rather, speaking or writing itself is a process that accomplishes thought and its meaning. Thought and the meaning of a thought do not happen apart from words, writes Merleau-Ponty, just as music and the "musical meaning of a sonata is inseparable from the sounds which are its vehicle (Merleau-Ponty 1962, 182). A person's speech is a person's thought (Merleau-Ponty 1962, 180). And written or said words (thought) are just one sort of behavioral process through which we express or accomplish a felt or imagined need. It is one of numerous human processes—each of which overlaps and integrates with the others. In Merleau-Ponty's eyes, then, Nietzsche's story, his story, and my story become meaningful in the telling of them, not before.

My story of Nietzsche's story about origins of felt and imagined experience pays special attention to his text *On the Genealogy of Morals* (1887/1989). I expose and interpret the ways Nietzsche makes sense of received categories of being, especially traditional dichotomies of being like mind and body, self and other, idea and matter, and psychology and biology. As has been widely recognized by commentators, Nietzsche receives these categories skeptically. Nietzsche's interpreters in the last three decades have especially emphasized themes in his texts about a loss of enlightenment concepts of reason, self, and identity. I offer a new interpretation. I contend that in his writings, Nietzsche not only refuses to settle with most traditional, often reductive or foundational, accounts of categories of being. Nietzsche also does the following two things. First, he points to three intersecting concepts at the non-foundational limit of his thought: dynamic non-dualism, relation, and metaphor. By dynamic non-dualism I mean a field of overlapping planes—especially the planes of the ideational, the psychosomatic, and the socio-physical. By relation I mean the being of relations without *relata*, that is, the being of relations without reified being. By metaphor I mean a sensation, image, or word that denotes an item as and in the place of another item. Second, through his nods to Heraclitus, Nietzsche begins to open his and our perceptual structures to modes of pre-literate human experience of which we are habitually unaware.

My view about Nietzsche then, shows him playfully advancing a non-official position characterized by three co-expanding and limiting concepts: dynamic non-dualism, relation, and metaphor, and cryptically suggesting that structures shaping perception since literacy are not the only ones pertinent or available to human speech and perception. It shows Nietzsche co-creating new bonds within and across traditionally perceived conceptual and cultural planes, all the while likewise co-creating his own felt and imagined life. As Nietzsche co-creates a story with reconfigured concepts about traditional categories, he lives his co-creation. He becomes, in an important and question-worthy way, his story. The following is an overview of the author's story about his.

PART II. OPENING NIETZSCHE'S GENEALOGY TO "FEMININE" BODY

Nietzsche's story in *On the Genealogy of Morals* (1887/1989) suggests that human beings produce a concept of what it means to be a human being at any given time in Western history according to a mutual shaping that occurs among the concept of the self and the developing constitutions of conscience and corporeal punishment. I choose these three constitutions[3]—self, conscience, and corporeal punishment—for a couple of reasons. They are themes in Nietzsche's *On the Genealogy of Morals* and they constitute both the extremes and center of a relational continuum I call Nietzsche's "dynamic non-dualism." In chapter 2, I provisionally categorize selfhood as ideational (the immaterial extreme), conscience as psychosomatic (the center), and corporeal punishment as socio-physical (the material extreme). My conception of Nietzsche's dynamic non-dualism displays the playfulness and co-creativeness that Nietzsche shows in his refusal to accept perceived "official" enlightenment stories about the boundaries of these categories. Thus, the borders of these categories—selfhood, conscience, and corporeal punishment—are not strict, but characteristic of a continuum, overlap.

In the formation of human beings' concept of self, a reciprocal shaping, suggests Nietzsche, occurs among the three constitutions: self (ideational), conscience (psychosomatic), and corporeal punishment (socio-physical). The way a culture establishes an interpretation and custom of corporeal punishment influences—and is influenced by—the constitutions of conscience and selfhood. Only after Western people perform a long labor of physical cruelty upon themselves, Nietzsche theorizes, does conscience develop a memory, making possible the idea of God, the "Platonic" fixed idea, and the idea of "eternal" guilt. This ongoing reciprocal movement within and across the planes of corporeal punishment, conscience, and human being's idea of self points to social influences participating in the formation of meaning and its physical manifestation in people's bodies. It points to a relationship between idea and body, and indicates that only provisional borders separate them. It points to an idea-body integration.

Because Nietzsche heightens the stature of body in the formation of the concept of self as implied above, he has been said to open Western philosophy to the other, the body. Even so, it has been said (Oliver 1955, 17–25), that although Nietzsche opens Western philosophy to the body, he does so to primarily a male body. In chapter 2, in addition to describing Nietzsche's conception of self formation as co-participant in a dynamic non-dualism, I create a framework—using a strategy that is itself Nietzschean—through which the possibility and questionability of a symbolically feminine body begins to emerge. I do this by using the metaphor of Indian curry. The metaphor works on two levels: as a symbolically feminine body; and as Nietzsche's conception of self-formation as a dynamic non-dualism.

The curry metaphor for Nietzsche's non-dualism reflects the illusory target that has been mistaken for a coherent "unified" self. Nietzsche's writing explicitly undoes the Enlightenment concept of the supposed unified self. In so doing it implicitly fractures the concept of the apparent natural woman[4] that presupposes a unified self. I argue that with respect to the concept of natural woman, Nietzsche's words say more than Nietzsche means. They stretch towards a feminist reappropriation.

The Indian curry metaphor symbolizes the incoherence of the supposed unified self as well as the feminine body. It ties together Nietzsche's non-dualism with the symbolically feminine body. The logic of Nietzsche's words, if not their symbolism, prompts one to build a bridge from his writing to a symbolically feminine body.

Additionally, many recent accounts of body by feminist philosophers aim to avoid a notion of body as biology while sustaining a non-Cartesian dualist conception of mind and body. An example of this is Moira Gatens's concept of body in *Imaginary Bodies: Ethics, Power and Corporeality* (1996). By "body" she does not mean biology but symbolic body. Although body as biology *only* is problematic for a nonreductive non-dualist (for whom what it means to be a human being can be reduced neither to the mental nor the biological), body as symbolic *only* also remains problematic. It reduces all to the mental.

My Nietzschean account of symbolic body attempts to avoid this reductivism. I categorize the images and metaphors that configure a symbolic body as ideational. And my concept of Nietzsche's dynamic non-dualism reveals that the ideational, by definition, is not purely idea. It overlaps planes with the psychosomatic and socio-physical. Thus, my concept of Nietzsche's dynamic non-dualism shows Nietzsche attempting to overcome the contradiction inherent in a supposedly nonreductive nondualist conception of body that nonetheless excludes the physiological.[5] It shows why physiology need not be excluded from a conception of feminine body. This approach makes symbolically feminine body a more convincing nonreductive non-dualism, and also could be extended to a symbolism of masculine body. Thus, it avoids suggesting that a woman—or by extension a man—is her or his biology only. It suggests that biology is not static, that women and men are not thereby limited to traditional Western roles.

PART III. NIETZSCHE'S ASCETIC IDEALS AND A PROCESS OF THE PRODUCTION EMBODIED MEANING

Nietzsche's story examines the plausibility of traditionally perceived identities—an identity of a concept, of a moral value, of a self, or of a semantic referent. It implies that the ancestry of his felt and imagined experiences—including felt and imagined reified identities—includes, conceivably, the absence of the human experience of such identities. I explain my interpretation of his story in part by my concept of dynamic non-dualism. Such a non-dualism depicts idea and matter, mind and body, self and other, and other traditional dichotomies as reciprocally related and engaged in a motion of exchange. Such dynamism underscores in Nietzsche's story a concept of self much like that which Lifton and Hisamatsu describe. It spills beyond the borders of any permanent traditionally hypostatized identity. Its inside and outside are not easily distinguished, in part, because the self is permeable to inner and outer influences. Moreover, just as Nietzsche's story of the concept of self is a story reminiscent of the Greek god of many forms, Proteus, so too is his story of body and other constitutions a story about mutual integration, fluidity, and change.

My view of Nietzsche's story about the concept of self-formation reveals a general pattern of exchange within and across the planes of ideational, psychosomatic, and socio-physical constitutions. In chapter 3 I extend my view to explore Nietzsche's story about the concept of "ascetic ideals" that Nietzsche talks about in his "Third Essay" (1887/1989). For Nietzsche, ascetic ideals (generally an ideational constitution) names a particular lineage and style of ideals emphasizing self-denial that Nietzsche inherits as a European—and especially as a European of Protestant descent. Nietzsche investigates the lineage of the concept of ascetic ideals. In his investigation he points to a paradox between multiple meanings for ascetic ideals co-created by humans and their environs, and an absolute meaning for them insisted upon by certain dominating "ascetic priests."

In chapter 2, in addition to the concept of dynamic non-dualism, I introduce the concept of relation. Through the concept of relation and, in particular, an etymological argument about the concept, I reveal in Nietzsche's writing a pattern of reciprocal relations and apparent absence of reified things. It may come as little surprise then that amid much of Nietzsche's ascetic-ideal discourse, the concept of ascetic ideals shows itself likewise indeterminate in meaning. It too exhibits a nonreified identity. In this context, ascetic ideals is a signifier signifying and interpreted in terms of many meanings, not a single univocal one. In *On the Genealogy of Morals*, Nietzsche reveals the importation of ascetic ideals into human existence as an attempt to fill a hole perceived to exist given the absence of absolute meaning for suffering. It distinguishes, in other words, an attempt to import into human experience proper meaning.

By proper meaning I mean both the identity of the meaning of a concept and the identity of the meaning of the being of beings per se. My criterion for proper meaning is that some form of permanent being exists and is an absolute measure and arbiter of truth. For instance, one criterion for proper meaning is that there would either be an aspect of permanence accorded phenomena or accorded their origin. Wittgenstein, in *Tractatus Logico-Philosophicus* (1961), distinguished between objects and propositions about objects. Objects do not in themselves involve the property of truth. Truth, if it exists at all, suggests Wittgenstein, presupposes the existence of objects. Truth, if it exists says Wittgenstein, pertains to assertions humans make about objects and states of affairs among objects. But, as Heraclitus intimates, truth presupposes in addition to objects and propositions about objects that there exists a condition of permanence either in the objects and states of affairs or in their origin (i.e., God, ideal forms, etc.). The classical Greek debate famously manifest in the contrasting views of Heraclitus and Parmenides illustrates the necessity of the existence of permanent being in order for their to be unconditional propositional truth. If indeed all being is constantly in flux, as some of Heraclitus's fragments suggest, then there seems to be no way to stay a proposition—to attribute it "truth." This is because the moment a proposition is stated, presumably the object and the perspective of the viewer in relation to the object will have changed. And so, if the possibility of "truth" is in question, so too is proper meaning—that is, so too is the identity of meaning of concepts and of the being of beings per se.

It may be the case that a realm of permanence does in fact exist and lend stability to our assertions about objects we experience. The existence of permanent being may be the case even if a person is not aware of it, let alone able to know or prove it. And many may assume or have faith that such a realm of permanence exists, despite not being able to ascertain such a realm. Many forms of Judaism, Christianity, Islam, and philosophy assume the permanence of being. Indeed, it remains an open question whether such permanent being or God exists. An aspect of Nietzsche's story tries to bring to visibility the habit of making such fragile assumptions as we go about our religious and/or technological practices.

Many of these assumptions about the existence of permanent being that would ground proper meaning have been absorbed into secular variants in the West of common sense. Henri Bergson (1908/1991) addresses one manner of secular assumptions about permanence and proper meaning: "But the separation between a thing and its environment cannot be absolutely definite and clear cut . . . the close solidarity which binds all the objects of the material universe, the perpetuality of their reciprocal actions and reactions, is sufficient to prove that they have not the precise limits which we attribute to them" (209). Bergson's questioning the discreteness of an object with respect to its environ simulates one of several ways that Nietzsche's essay about ascetic ideals brings into view the apparent absence of invariance and proper meaning. We often assume that solidity and permanent meaning exist regarding particular objects in the world. One object for us is the concept of human being. Ascetic ideals especially concern ideals ascribing ultimate or proper meaning to human existence. Nietzsche focuses on showing the seeming absence of permanent limits to the concept of ascetic ideals—a concept which itself concerns the question of ultimate or proper meaning of being. Thus, the chapter's interrogatory title—"What is the Meaning of Ascetic Ideals?"—might also be worded "What is the Meaning of Meaning?"

The question of the meaning of ascetic ideals is especially abstract. It not only asks about the meaning of an abstract *constitution* (i.e., ascetic ideals or meaning), but, one could argue, it asks the meaning of a particularly abstract constitution: the meaning of *meaning* itself. Because of the especially abstract nature of the latter, I begin with an apparently less difficult topic to illustrate: a kind of reciprocation characteristic of a process of the production of meaning in Nietzsche's concept of ascetic ideals. This first model will address the question framed by the phrase: "What is the meaning of (pre)menses?" I choose this phrase of the same structure as "What is the Meaning of Ascetic Ideals" because the structure of such genitive phrases harbors in itself a story that complements Nietzsche's story. I am working with genitive phrases in which the preposition "of" is immediately preceded and succeeded by a noun. This is the case for the following three genitive phrases: "the meaning of ascetic ideals"; "the meaning of meaning"; "the meaning of (pre)menses." By first exploring the genitive phrase "What is the meaning of (pre)menses," I provide a grammar that obscures conventional designations of subject and object. This is because each term preceding and succeeding the "of" acts as both subject and object. I compare this basic vision to a more developed co-constituting process among moral and definitive valuations of embodied experiences of (pre)menses. I show that *as* co-constituting, these factors—moral valuation, definitive valuation, and the (pre)menses constitution—are relations without *relata*.

As a "disease," premenstrual syndrome has received much attention in recent decades and some include it among a category of illnesses called "socially-constructed disease." Anthropologists, physicians, and even some philosophers are debating the status of its reality and its symptoms in relation to its primary cultural breeding ground: industrialized communities. My aim is not to take sides in this debate but to use the (pre)menses example to illustrate a process of the production of meaning implied in Nietzsche's concept of ascetic ideals.

Using data from clinical studies of women's experience of (pre)menses, I show that the way a culture establishes an interpretation of (pre)menses affects the way a woman physically experiences (pre)menstruation. In the West a predominantly negative valuation of menses inscribes itself both psychologically and physically in a woman's body. This bodily inscription is reflected back to the community and confirms and further fuels the earlier pejorative valuation. The formation of that which some call a socially-constructed "disease"—premenstrual syndrome—begins to emerge. This process models the reciprocal movement in *On the Genealogy of Morals* that I discuss in chapter 4. It registers a culture's valuation of a constitution and the individual's received and reciprocating conceptualization and embodiment of that value. This general pattern that I have developed based on Nietzsche's "Second Essay" reveals, perhaps more simply when viewed through the (pre)menses example, one of the reciprocal movements characteristic of what I have been calling Nietzsche's dynamic non-dualism. One can begin to see in this reciprocal movement and dynamic non-dualism how Nietzsche's story exposes an alternative to the legacy of Cartesian dualism and Enlightenment ideas of self, made possible by presupposing the being of proper concepts.

And so, by using the example of (pre)menses, I lend support for Nietzsche's way of saying if not explaining the often-confusing relationships and experiences we feel and imagine. Nietzsche's story, I believe, suggests that what we have tended to call respectively our mind and our body we experience as a relation without related things, that is—without discrete immaterial "mind" or discrete material "body." If the (pre)menses example serves to illustrate Nietzsche's story, then it seems Nietzsche's story reciprocates by serving to expose the contemporary phenomenon of (pre)menses. Nietzsche's story when applied to the (pre)menses example, indicates that the experience of (pre)menses is generally not a static function. In other words, when we turn from the genitive phrase, "What is the meaning of (pre)menses?" to the genitive phrase "What is the meaning of ascetic ideals?" my view of Nietzsche's story shows that a process of the production of meaning does not involve or produce a static function of reifiable identities.

The (pre)menses example shows that how women experience their bodies during the premenstrum is in part a function of how a society values women. It shows in other words, that that which we discover the premenstrum to be is in part a function of the role, status, and value we discover in or assign to women. Whether the pattern I develop in this model can be extended to the historical and social genesis of other apparently socially constructed diseases remains an open question, but the pattern itself is reproducible to that end.

PART IV. NIETZSCHE'S ASCETIC IDEALS AS A PROCESS OF THE PRODUCTION OF MEANING

If extending the above model to certain other apparent bodily ills seems an open question, extending it to other concepts is not. In chapter 4 I show how the above model can help explain the genesis of one concept on which Nietzsche focuses: the concept of ascetic ideals. The general pattern of the (pre)menses example helps to show a reciprocal movement characteristic of the dynamic non-dualism I attribute generally to Nietzsche's writing and especially to his *On the Genealogy of Morals*. It shows the reciprocity via a seemingly more "concrete" object of a genitive phrase, (pre)menses. In chapter 4 I show the same pattern operative through a seemingly more "abstract" genitive object: the concept of ascetic ideals. In both situations this pattern of a reciprocal movement expresses a culture's valuation of a constitution and the individual's conceptualization and embodiment of that value. How a woman experiences menses, or how a person experiences the concept of ascetic ideals—especially a semantic referent ascribed to it—will probably never correspond, Nietzsche's story suggests—with a static and once-and-for-all identifiable entity. The process according to which phenomena arise for humans indicates instead that whether largely embodied ([pre]menses) or largely conceptual (the concept of ascetic ideals), phenomena lack reified existence. Phenomena seem to rise, vary, alter, or pass away according to a steady flow of complex factors.

My view of Nietzsche's dynamic non-dualism shows the elastic and fragile origin of the concept of ascetic ideals. It implies likewise that the origin of any concept is probably like the origin of the concept of ascetic ideals. It is not—nor possesses ancestry in—an Origin. Concepts and their meanings, my story about Nietzsche's says, originate not from a firm unchanging foundation (like Plato's Ideas) but from a continual flow and exchange of complex factors. These factors fuel and constitute Nietzsche's dynamic non-dualism. Through what I call Nietzsche's dynamic non-dualism, Nietzsche's story exposes an alternative to the legacy of Cartesian dualism and Enlightenment ideas of self. The Cartesian and

Enlightenment legacy has often been presumed possible by assuming the being of reified concepts—grounded in a firm Origin. Nietzsche's story muddles the plot of this legacy. It peels away layers of genealogical flesh that have kept hidden from us, fruits intimating an apparent absence of reified concepts.

Nietzsche's story peels back the skins of history. As skins are pulled back with regard to the concept he has come to call ascetic ideals and that has much to do with practices of self-denial, Nietzsche's tale reveals for us a complex genesis. Although it seems that we all tell stories, the telling of any singular story is not, Nietzsche implies, necessary for any one of us. And yet we often cling to our stories. We often assume our stories about our pasts are necessary fixtures of who we are now, of how we might be heard and understood by others, and of who we imagine we might become. It is unconventional for people living in mainstream Western societies to believe that none of the attributes of these personal stories need be attached to us. Just because we experience an event, and perhaps integrate it into our sense of who we "are" or have become, does not mean, Nietzsche suggests, that we are—or need be—defined even partially by the event.[6]

PART V. NIETZSCHE ON A PRACTICE AND CONCEPT OF GUILT

Many of us in techno-rational societies try to live up to self-imposed ideals, ideals we view as proper ways of comportment. Even if they are so subtly ingrained that they flow from us prereflectively—like habits—we try to live, it seems, by certain precepts. Many of our precepts have been influenced by certain Christian or American-Puritan ethics. "Don't be a burden" is a precept some of us tell ourselves, simultaneously discouraging interdependence and encouraging self-reliance. "Eat and drink in moderation" might be another, as well as "Keep your word." These three precepts arguably have roots in the Plato-Christo-Protestant origins of much of Western culture. The precepts to which we hold ourselves, like mullions bending light through windowed glass, alter our self-seeing and seeing of others. When we believe we have fallen short of our internalized precepts and self-expectations—a common experience for selves in the West—our perceived inability grows up as a feeling of guilt. Chapter 5 explores Nietzsche's writings as a critique of a concept of guilt.

We lean strongly, Nietzsche shows, into a view of the feeling (and concept) of guilt as a built-in human condition. In the nineteenth and twentieth centuries, however, some began to query this assumption. Herbert Fingarette, for instance, explores the practice and texts of ancient Confucianism and tells us that guilt is an especially Anglo-European behavior and concept. Among early Confucian-Chinese societies, he alleges, there is a concept and feeling of *shame* but not guilt (Fingarette 1972). Whereas shame implies a public and externalized sense of wrongdoing—usually relating to one's family or country—guilt, explains Fingarette, implies a private and internalized one. According to Fingarette, the Confucian of ancient China has almost no internalized sense of self nor by extension, guilt. One's identity, strengths, and shortcomings commingle necessarily with an externalizing "self" that involves more than one individual—a "self" (more like what we understand as family) that extends to and includes one's brothers, sisters, mother and father, grandparents—and encompasses by degree and proportionately lesser potency, neighbors, and nation-state.

In her visits with the Dalai Lama, American teacher of Buddhism Sharon Salzberg asked the Dalai Lama for advice.[7] How does he teach his students to show *metta* (the Pali word often translated "loving-kindness") for themselves and mitigate the suffering that comes from attitudes of self-loathing and feelings of worthlessness and guilt? At first the Dalai Lama doubted the importance of the question. He told Salzberg that such self-loathing is probably particular to her, implying that feelings of personal guilt are not a common form of human suffering. Indeed, he had not given the topic much thought. The Dalai Lama had difficulty fathoming it a widespread experience among any society. Salzberg continued to press him on the question, insisting that many people in the United States suffer from feelings of unworthiness. Still incredulous, the Dalai Lama proceeded to go around the room asking each person of Western background if they and those they know had had such experiences. Each said yes. He concluded that this is a great difference between his Tibetan society and those of Western descent. It seems, that based on this tale, suffering from feelings of unworthiness and personal guilt may not be a significant aspect of the lives and practices of people in all societies.

Nietzsche's attitude toward the concept and practice of guilt like Fingarette's and the Dalai Lama's, invites folks saturated in Western Judaic, Christian, Muslim, or American-Puritan precepts to imagine the possibility of human lives without guilt. By questioning traditionally perceived conceptual and felt limits, Nietzsche's story is one of several participants in this invitation.

We selves, Nietzsche tells us, seem to be immediate fields seeded and sowing permeable planes—planes traditionally (and misleadingly) understood as discrete, and as bifurcating selves and others, subjects and objects. Nietzsche's story anticipates and even breathes into other nontraditional stories by Bergson (1908/1991), Heidegger (1927/1996), and Merleau-Ponty (1945/1962), to name a few—bridges across human felt and imagined worlds. Nietzsche's nontraditional story with respect to a dominant Western practice and concept of guilt, like to that of ascetic ideals—shows the production of such a practice and concept happening according to the concepts of dynamic non-dualism and relation.

At the beginning of my introduction, I described Nietzsche's storytelling method in terms of Nietzsche positioning competing substories. Nietzsche, I wrote, rarely weaves a substory without also weakening the story by challenging it. Among these substory positions, none is an "official" position. And yet, in telling Nietzsche's story thus far, I have emphasized two concepts: dynamic non-dualism and relation. If one contends that dynamic non-dualism and relation consistently guide Nietzsche's story, does one not suggest that Nietzsche's story does offer an "official story," if not a traditional one? Is one's view of Nietzsche's, in the end, contradictory?

PART VI. NIETZSCHE, METAPHOR, AND BODY

Does Nietzsche offer in the final analysis not so much a bunch of mutually mobilizing stories, but instead a privileged story among them—one that settles his view of the way things are such that the concept of dynamic non-dualism and relation becomes a sedimentation that *is* Nietzsche's story of "reality"? Chapter 6 answers these questions by turning to Nietzsche's 1873 text, "On Truth and Lying in a Non-Moral Sense" (1999), which becomes pivotal for the argument. It is by 1873 that Nietzsche has shed the residue of Schopenhauerian metaphysics—of 'correct perception' and its cause—arguably present in his *The Birth of Tragedy* (1872/1967). As critical to this part of Nietzsche's story is an aspect of Nietzsche's tale-telling revealed by Eric Blondel.

According to Blondel, Nietzsche's writings indicate that the body constitutes reality as it interprets it. For Nietzsche the bodily drives do not reproduce an object. They include simultaneous capacities to constitute and interpret. These dual capacities to constitute and interpret, co-create, and comport the very objects of our experience. I contend that for Nietzsche the body's status as interpretation is made possible in part "as metaphor and through metaphor."[8] Indeed, the concepts of dynamic non-dualism and relation respectively could in this light be said to be instances of metaphor. I emphasize like Blondel and at lease one recent Nietzsche interpreter, Tim Murphy,[9] that the concept of metaphor works for Nietzsche as a (non)foundational basis of language, concepts, and perception.

One view opened up by the concept of *metaphor* in Nietzsche's writings is Aristotle's definition of metaphor. Aristotle writes in *Poetics* (trans. 1941) that the concept is logically "proper" and the metaphor "improper." Metaphors, says Aristotle, compare items that belong to one category of being to items housed in a completely different category. Because metaphor involves carrying an item out of its allegedly proper sphere into an improper sphere, metaphor is logically improper, suggests Aristotle.

A view like Aristotle's emerges in "On Truth and Lying in a Non-Moral Sense" when Nietzsche describes human sensory and intellectual experience as a multiplicity of translations or transferences from one sphere to another. There is no right perception of an "item"—no proper object nor any proper sphere. There are, rather, permutations of constellations of forces as they move within and across arenas. Here I consider any conceivable part of a whole—or whole "itself"—an arena. All phenomena could be said to be metaphors in the sense that they are translations of or stand-ins for an item. The item may have been otherwise because of the effect of memory and the passage of time regarding the item/metaphor, as memory translates the item/metaphor from one moment to the next *within* an arena; or it may be otherwise because it is being perceived *across* arenas, let us say to the arena of smell, and thus is translated by the nose into a scent.

For instance, I might come across a hillside of grey-green sagebrush. I smell its scent. It is a fragrance too complex to generalize by words like "pungent," "cleansing" and "astringent." My perception of the sagebrush fragrance does not capture any aspect of the sagebrush in a proper sense. The scent I perceive is more like a translation of, or metaphor for, certain sagebrush forces. The bitter-fresh scent may in turn be translated again into another metaphor as it becomes a memory for me—a simplified brand of the initial smelling experience and of an entirely different composition. I might later discuss the idea of "sagebrush" in a botany class and conjure in my mind the abstract concept "sagebrush." For me and other students respectively, the abstract concept "sagebrush" stands again as a metaphor—this time further removed from one's experience of particular sagebrush bushes than the first two. In each of the sagebrush phenomena above—in the smelling, the scent remembering, and the sagebrush conceptualizing—each can be called a "metaphor." Each carries over, literally *meta-phors* or *trans-lates* from one time or arena into an entirely new time or arena.

The sagebrush example suggests the absence of any proper human experience of sagebrush. It illustrates the sort of anthropomorphizing of forces that Nietzsche discusses in "On Truth and Lying in a Non-Moral Sense." Another living being, like a bird or a tree, Nietzsche writes, will not share a like translation of the sagebrush, but will create and share among its kin like translations or metaphors of them. These shared metaphors are not identical but more similar within than across species.

In this part of Nietzsche's story emphasizing metaphor and conceptual propriety, proper meaning appears—as in parts described earlier—not to exist for humans. Nietzsche's apparent elevating of the improper over the proper, this time, in the form of the metaphor over the concept—parallels an aspect of what I call relations. It also overlaps with an aspect of the concept of dynamic non-dualism. It does so through a new linguistic and conceptual prism: through the language and concept of metaphor. Whether using the language or concept of metaphors, relations, or dynamic non-dualism, in all three cases one names constellations of forces that are identifiable only provisionally. With respect to their "objects" respectively, a metaphor, a relation, and a dynamic non-dualism is not proper but protean and otherwise-ing.

My argument emphasizing the concept of metaphor in Nietzsche's writings adds new folds to this shared protean fabric. It shows Nietzsche's metaphors deliberately blurring a distinction between mind and body. It shows cognitive imagery and language typically associated with intellection configuring Nietzsche's descriptions of physiology, and diction often linked with the body illustrating the intellect. One effect of Nietzsche's switching the traditional attendant imagery and language is an obscuring of a perceived boundary between body and intellect. Another effect is that a switch in priority takes place. Because the language traditionally associated with intellect has typically been more positive than that linked to the body, in Nietzsche's accounts there is, conversely, an elevation of body over intellect. The concept of body now has in its territory the flattery usually attending the concept of intellect. Vice versa, adjectives such as "obtuse," "dumb," "irrational," and "simplistic"—typically attending descriptions of the body—cluster around the intellect. A reversal of values occurs. The body is both conflated with (if not entirely) cognition *and* playfully and strategically elevated above it. As Nietzsche conflates body and mind, he distinguishes them too, in part by interposing the body between the chaotic forces of the external world and the simplified concepts of the internal mind. My view is similar to Blondel's on this point. For Nietzsche, the body unites the mind and impinging forces of the outside world into an interpenetrating non-dualism according to which a mixing, but not a reductive merging, of difference occurs. Nietzsche's mix tries to interconnect, but not reduce, body to mind or mind to body. It tries to avoid a reductive material or spiritual monism while at the same time trying—via the hypothesized mix—to avoid a Cartesian-like dualism.

If for Blondel, Kofman and Murphy, however, the concept of metaphor names the primary way, if (non)foundational, to understand body as interpretation for Nietzsche—that is, to understand human perceptive and linguistic experiences—my view differs in the following way. It steers away from an implied reduction of Nietzsche's philosophy to the concept of metaphor. That is to say, it avoids understanding itself in terms of a single if (non)foundational basis. "As and through metaphor," I contend, articulates only one way that Nietzsche suggests humans can say and experience their (non)foundational status as beings. My interpretation of Nietzsche's story points to at least two others. It points to the concept of dynamic non-dualism and it points to the concept of relation. These three concepts—dynamic non-dualism, relation, and metaphor—say "the same" idea. According to my position, the same idea however does not exist determinately. Each of the concepts describes this same idea through a different argument.

My interpretation of Nietzsche's story could not be articulated independently of the three concepts and their respective arguments, language and style. This is not to say that these three concepts are the only ones through which this same idea might be articulated. My argument divulges, perceives, and co-constitutes the concepts of dynamic non-dualism, relation, and metaphor with Nietzsche's text. It exists *as* and *through* these three variegated concepts. It exhibits itself through their splayed sameness, and through their implied interstices and spontaneous ruptures in Nietzsche's writing.

The concepts of dynamic non-dualism, relatedness, and metaphor, then, are the same in that they say the same idea and different because they speak different arguments and styles. Specifically, the concepts are the same for these reasons: each involves an array of origins but not a proper origin; and each reveals impermanence and change as persisting factors of force configurations. The arguments of the concepts are different in these ways: Nietzsche's story says the concept of dynamic non-dualism in terms of a triad of overlapping planes; one ideational, one psychosomatic, and one socio-physical; it says the concept of relation in terms of a buried etymological argument of the German word *Verwandtes*; it says the concept of metaphor in terms of human perception, language and concepts operating from metaphor as their (non)foundational foundation. The three concepts, then, are both the same and different.

For Nietzsche, I show in chapters 2, 3, 4, and 5, that there appear to be no proper concepts. If one applies the idea of the absence of proper concepts to the concept of dynamic non-dualism, it means that the concept of dynamic non-dualism is the same as itself and yet not identical to itself. This is because the concept of dynamic non-dualism appears to have no persisting identity. In a weakened but parallel manner to the self-relation of a concept, the concepts of dynamic non-dualism, relation, and metaphor can be said to be similar but not identical to one another.

Moreover, each of the three concepts names a (non)foundational basis that can be said to permeate, flexibly organize, and disrupt a centralized ordering of all three by any one. This works like a balance of power. Each of the concepts configures an argument that can be said to tie together (in its own manner) all three of the concepts. The concept of dynamic non-dualism displays an architecture that is (non)foundational for Nietzsche's story. Its tri-planed architecture provides geometrical strategy for conceptualizing Nietzsche's undermining of Cartesian dualism. It provides a spatially oriented structure with which to associate and envision the concepts of metaphor and relation. The concept of relation shines light on the rich entanglements of Nietzsche's words—their etymologies serving as a wealth of excavated goods for showing the historically qualified origins of apparent Origins often perceived as unqualified. It provides a thesis about the (non)foundational ancestry of a word: To associate this thesis with the concepts of dynamic non-dualism and metaphor lends an added vantage point from which to consider the arguments displaying the other two concepts. Indeed they too are scripted in words with etymologies even if they are not ostensibly about words and etymologies. Likewise, the concept of metaphor expands the set of angles for beholding and living the concepts of dynamic non-dualism and relation. New vistas emerge when aspects of the concepts of dynamic non-dualism and relation are considered both in terms of certain metaphors I produce to symbolize them, and in terms of the layers of "meta-phor" or "trans-lation" constituting in general a subject's perception of objects, and in particular a reader's perception of my argument about Nietzsche' story.

Nietzsche's story of interpretation locates the concept of metaphor alongside those of dynamic non-dualism and relation. Each of the concepts appears to be similarly basic and (non)foundational in Nietzsche's story. Each expands, complements, and challenges the organizing logic of the others. Each remains permeable to the forces of the others. My argument about the concept of metaphor in Nietzsche's writings differs from those of Blondel, Kofman, and Murphy because it views in it the described limited stature. The concept of metaphor competes among other possible (non)foundational foundations in Nietzsche's writing. My interpretation of Nietzsche's story plies out the possibilities, bringing three into relief. In a position to answer the chapter's opening questions, I show why asserting the (non)foundational intersection where the concepts of metaphor, dynamic non-dualism, and relation cross as Nietzsche's official story about the "real" recoils in at least three ways.

PART VII. NIETZSCHE AFTER NIETZSCHE

If Nietzsche's writings stand out for their wariness of an official story, they nonetheless stand with at least some scholarly and artistic inquiry in the twentieth and twenty-first centuries in their exploration of human experience as dynamically non-dual. Indeed, in the twentieth and twenty-first centuries, Nietzsche's story can be said to be less a submerged story than in preceding centuries.

Like the writings of Nietzsche, the writings of phenomenologists Edmund Husserl, Martin Heidegger, Jean-Paul Sartre, Maurice Merleau-Ponty, and Emmanual Levinas have been recognized by many as providing alternatives to a Cartesian-dualist and Enlightenment-subjectivity worldview. If Nietzsche's response to Cartesian dualism, enlightenment subjectivity (i.e., Kant), reductive materialism (i.e., Marx), and reductive idealism (i.e., Hegel) is not the only nineteenth-century response, it is one of the most effective. In the purview of twentieth-century phenomenology, the same might be said of the writing of Maurice Merleau-Ponty. For this reason, chapter 7 extends the project beyond Nietzsche to the ways Merleau-Ponty approaches some of Nietzsche's concerns.

A phenomenological approach[10] provides one of the most promising ways to contend with inquiries of Nietzsche. Methods of phenomenologists are more structured than those of Nietzsche. Although I do not want to suggest similarity between Nietzsche's methodology and that of phenomenologists, striking commonalities do exist among some of their assumptions, themes, and emerging insights.

Two such similarities regard the concepts of "concept," "self," and "origin." Merleau-Ponty, for instance, suggests an absence of a proper limit to any of our felt or imagined experiences of each. Like Nietzsche, he tells of beginnings, but not *a* beginning; of appearances of concepts—including the concept of self—but appearances whose borders shift and fade away into a horizon imperceptibly. And although it would be wrong to attribute to Merleau-Ponty's story the dynamic non-dualism, relationality, and metaphor I attribute to Nietzsche's, a comparable concept emerges in Merleau-Ponty's writings. This is the concept of *Gestalt*. According to the concept of *Gestalt*, human experience is a whole, not dual (mind and body). And as a whole, it unites mind-body-world while maintaining multiplicity and singularity. It unites them in a story that uses the language of "intention," "motivation," "direction," "desire," "sexuality," "expression," and "thought" to describe elastic and intertwining human behaviors. Using a more visibly structured method than Nietzsche's story, Merleau-Ponty's story invites comparison to Nietzsche's. It dynamically unites—or tries to—traditional dichotomies of mind and body, interiority and exteriority, and subject and object.

There has been increasing attention to the writings of Merleau-Ponty in the last two decades. I emphasize themes of trauma with regard to corporeal adaptation. Few have placed Merleau-Ponty's work in conversation with contemporary trauma studies,[11] or acknowledged a concept of trauma implicit in Merleau-Ponty's writings.[12] I show Merleau-Ponty's thought illuminating ordinary experience through an examination of the extraordinary—traumatic experience. What becomes visible is corporeity existing as communication[13] that varies from primordial to complex; and that such communication is rooted in corporeity that directs itself outward to delineate itself and make of itself a presentation; and elasticity and interpenetration across the overlapping planes of human behaviors.

Merleau-Ponty appeals to the case studies of brain-injured men, and amputees of World War I[14]—survivors of "physical" and "psychological" trauma. If Merleau-Ponty implies the exceptional status of trauma for his project, he nonetheless leaves virtually unexamined the concept of trauma. In chapter 7, I begin such an examination. I specify three kinds of bodily adaptation—anatomical, technological, and verbal—and show each as expressions of fulfillments of preconscious desire. My analysis looks especially at these three modes of adaptation when they are responses to trauma, that is, when they are in the throes of living and adapting to human being's most demanding kind of preconscious desire.

Two kinds of bodily adaptation are anatomical and technological adaptation. By anatomical adaptation I mean the prereflective bodily process according to which organs, tissue, veins, and other aspects of anatomy shift to accommodate bodily needs stimulated by changing conditions (i.e., a developing callous on a part of the body newly exposed to repeated pressure or friction). By technological adaptation I mean a dually reflective *and* bodily process. Our body exercises reflection *and* built-technology in order to accommodate bodily needs that are not or cannot be met by anatomical shifting. One would first reflect and "make-up"[15] an adaptive apparatus to meet a felt need (i.e., imagine a chair to relieve the pain of one's feet) and then actually build the apparatus (build the chair to actually relieve the pain).

Important here is the way corporeal responsiveness reveals anatomical and technological adaptations to exist as communication. In the case of anatomical adaptation, I examine processes of spinal nerve systems after spinal cord trauma. Complex communication (learning and memory), traditionally presumed exclusive to the more complex cerebral neurons above the spinal cord, has been shown in recent studies of spinal-cord trauma to occur below the cerebrum. Attempting to survive the new circumstances after an injury, spinal neurons vigorously interrogate, respond to and learn from their new conditions. The nerve processes of a spinal-cord injured body can be seen reorienting themselves and *creating* new strategies for survival in the altered conditions. This is significant because it challenges traditional views that separate bodies from minds by allowing for complex communication—traditionally associated with mental and cerebral activity—to happen below the cerebrum. The disclosure of such creative activity independent of the cerebral cortex allows us to "return" language to the "place" it has always been: our bodies and nature.

If some felt needs are successfully met by anatomical responses to them, others are not. In these circumstances, corporeity often responds by building technologies for a person with a certain disability in order to meet a need typically managed independently by persons without the disability. Such technologies—a prosthesis or wheelchair for example—extend our lived bodies beyond our anatomies. If technological adaptation does not, as anatomical adaptation, return communication to the interior organs of anatomical body, it re-orients and newly demarcates the exterior surface of body as a signifying symbol. By revealing that corporeality is a self-designated self-showing intending itself to be seen by others as a symbol of oneself, we can see that a body that accrues assistive technology is compelled to re-structure itself for itself and others as a self-designated self-showing. The broad significance of technological behavior is similar to that in the example of anatomical behavior. It shows communication operating not merely above the spinal cord but amidst the lived body, thereby returning language to bodies and nature.

A third kind of bodily adaptation is verbal adaptation. A mode of verbal adaptation of speech that I explore is narrative storytelling. Merleau-Ponty only implicitly indicates the role such storytelling might have in expressing or fulfilling a preconscious desire (Merleau-Ponty 1962, 177–193). To lend informed explicitness to Merleau-Ponty's study, I consider Judith Herman's *Trauma and Recovery: The Aftermath of Violence–From Domestic Abuse to Political Terror* (1997). Judith Herman is a clinical psychiatrist. Her case studies focus almost exclusively on male survivors of combat and female survivors of rape. Combat and rape trauma, Herman writes, are the two most prevalent forms of trauma for men and women respectively in the twentieth century. In *Trauma and Recovery*, she documents combat and rape survivors reintegrating and reconstructing traumatized memory through speech.

Merleau-Ponty differs from Herman in that he does not emphasize recovering through storytelling. My argument shows that Merleau-Ponty and Herman's analyses fill lacunae in one another's projects. For both Merleau-Ponty and Herman, human experience of trauma creates an overwhelming preconscious desire in the survivor. Moreover, for Merleau-Ponty, speech or thought shows itself, like other behavioral processes (phantom limbs, anatomical adaptations, technological adaptations), to be born of preconscious desire. Speech is an additional way corporeity expresses such desire. The work of both Merleau-Ponty and Herman shows how bodily adaptation in the mode of speech, can express bodily desire and create a sense of meaning, the roots of which spring from prereflective yearning absent of the eventual meaning. In comparison to the two previously discussed adaptive modes, verbal adaptation points not so much to a process of communication below the cerebrum, nor to a technologically extended corporeity redemarcating one's surface symbolism of self. Rather, verbal adaptation points to one's reflective experience—and correspondingly, one's relationship to others—becoming restructured. This can involve reassessing one's religious or philosophical convictions and compel one to a new organization of one's metaphysical views and self-understanding. Changes in one's self-organization reverberate out, often requiring whole-scale restructuring of one's relationships with family members, friends, and social communities. Verbal adaptation does not dissolve extreme preconscious desire born of traumatic experience; it appears instead to be preconscious-desire inspired expression that can lead to a pragmatic reorganization of oneself. The significance of verbal adaptation mirrors that of anatomical and technological adaptation in two important ways. First, verbal adaptation exhibits one's reflective life tied to bodily desire, showing it rooted there. Second, having returned language (here understood as reflection) to the body, an analysis of verbal adaptation also shows the body "returned" to a natural environment that includes a community of others.

In addition to describing verbal, anatomical, and technological adaptation, chapter 7 proposes a more general thesis: that trauma for Merleau-Ponty operates as a magnifying extreme that makes visible, dynamic adaptations of our verbal, anatomical, and technological bodily processes. These processes show lived body *as* an intended self-showing rooted in and displaying bodily desire. Finally, that each of the explored bodily processes (speech, anatomy, and assistive technology) participates in a shared aim of self- and other designation, also underlines Merleau-Ponty's view of human experience as a dynamic whole. In this sense, Merleau-Ponty's thought can be compared to Nietzsche's. It too suggests an experience of self and world as non-dual.

PART VIII. NIETZSCHE BEFORE NIETZSCHE

Not only after Nietzsche's death, but before his birth, stories akin to Nietzsche's appear. In *Twilight of the Idols* (1889/1968b), Nietzsche singles out the early Greek philosopher Heraclitus. Heraclitus is one of the few philosophers in the West, says Nietzsche, deserving praise. Just what is the significance of Nietzsche's favoring Heraclitus? And just how does its significance bear upon one's view of the concepts of dynamic non-dualism, relation, and metaphor in Nietzsche's story? In chapter 8 I consider how the concept of dynamic non-dualism could be said to relate to the sayings of Heraclitus. Nietzsche offers only a few cryptic comments about why he respects Heraclitus so. Because Nietzsche offers little actual analysis of the writings of Heraclitus, I root my interpretation of Nietzsche's attitude toward Heraclitus in my interpretation of select Heraclitean fragments.

In certain fragments of Heraclitus, one can find a story different from that which Nietzsche calls "Socratism."[16] Nietzsche notes that there is a tendency in the European cultural and intellectual tradition to privilege reflective intellection over prereflective physiology. This tendency is generally bad aesthetics, Nietzsche writes, and can be seen emerging with Socrates. Nietzsche attributes to Socrates the unleashing of this "poor taste" and calls it Socratism. It was Socrates, Nietzsche writes, who emphasized that that which is beautiful must also be intelligible. It was Socrates who launched this basic premise of Socratism, Nietzsche alleges in *The Birth of Tragedy*.

My interpretation of Heraclitus's story shares similarities with my view of Nietzsche's story. It shows that Heraclitus's fragments help Nietzsche to fend off Socratism's associations of the beautiful with intelligibility. Heraclitus's fragments can be said to undermine this proto-modern aesthetic via three preliterate perceptual structures visible in the fragments. The first shows the possibility of the existence of human prephonetic speech and writing to depend upon the animal creatures, vegetative life, and geo-elements of the natural sensuous environment. This is significant because it suggests that the possibility of abstract thought—often understood in Christo-Platonist dominant cultures as the human capacity that renders human life separable from other forms of earthly life and from the human body as well—also depends upon a "more-than-human world."[17]

The second preliterate perceptual structure visible in several Heraclitean fragments, points to, like Nietzsche's writings, an obscured boundary for distinguishing subject and object. A conflating of subject and object happens in those of Heraclitus's sentences which dispense with a subject-object construction. Such sentences are unlike most in the West in the last 2,500 years—sentences in which the structure generally implies the existence of a subject and an object. Some of Heraclitus's sentences, by contrast, are structured by a verb tense of ancient Greek that specifically obscures the distinction of agent and patient and tenses the event midway between subject and object. This Greek-verb tense is called the middle voice.[18]

The third and perhaps most striking similarity between Nietzsche's and Heraclitus's speech respectively is the implied absence of a proper origin and permanent self. Although the two preliterate perceptual structures above implicitly point to such an absence, Heraclitus's concept of *psyche* explicitly does. It indicates a human perception of self that has not yet made the eventual transition from an indeterminate "Homeric" to a sedimenting "Platonist" self. A perception of *psyche* (soul or self) that some say persists from the time of Homer to roughly 500 B.C.E.[19] is not yet viewed as having a so-called "proper" conceptuality and origin. The concept of *psyche* has not yet come to be experienced as persisting, identifiable, and rational—attributes emerging especially with the reception and preferred interpretations of the writings of Plato and Aristotle. The traditional view that such attributes do attend one's self-understanding is not to be confused with the writings of Plato and Aristotle per se.

Each of the characteristics I enumerate above—(1) a perceptual existence that depends upon a "more-than-human" sensuous world; (2) sentence structures that obscure traditional boundaries of subject and object; and (3) an indeterminate concept of self—possesses characteristics akin to those implied by concepts of dynamic non-dualism, relation, and metaphor. They show kinship between Nietzsche's story and Heraclitus's. Such kinship may explain Nietzsche's affection for Heraclitus. It shows that well before Nietzsche, a story with resemblance to his appears. Most important, Nietzsche's nod to Heraclitus's speech opens his and our arguably constricted speech and perceptual structures to possible understandings and opportunities from which we could benefit and of which we are habitually unaware.

Thus, although Heraclitus and Nietzsche's styles are anything but similar, some of their remains appear as next-of-kin. Notable is their shared sensibility about origins and human selves. For both, human origins appear for us as no Origin. We humans, they suggest, seem to continually begin again, renewing and reshaping origins that, however freshly present, forge a trail of complex ancestry. And as we start, it would seem, we have choices. We can choose to tell cousin-stories about starts. My examination linking Nietzsche, Heraclitus, and Merleau-Ponty chooses such a start. It indicates that we can choose whether to tell versions of official stories or unofficial ones—of ones that anchor our renewed origins in an Origin, or ones that do not. We can choose to frame our stories from the middle of a cultural stream, or from the water's edge. Whether we choose to proceed from the one or the other, is any one of our telling stories in the end necessary? If not, should this dissuade our telling them?

It has been said that the telling of our stories is not only important but self-defining. What we say about ourselves and our world shapes the texture of who we are. For those of us living in a society aware of its own modern and postmodern situatedness, a human Origin is often in question. The meaning in our lives, some would say, seems primarily to emerge from that which we ascribe it and create. It has been said that, in a related sense, we become our telling ascriptions.

And yet, if we become our stories, do we un-become them too? If no story is necessary, then we *are not*—in any precise sense—any one of our stories. If we wanted to stop telling a story rooted in certain personal memories, could one? And if one could, would one become accordingly otherwise? My analysis of Nietzsche's story says yes and yet, not so easily. Nietzsche's story says an *agon*—a struggle. We are free if by "unfree" we mean unable once and forever to change certain *essentia*. We are unfree if by "free" we mean "the foolish demand to change one's essentia arbitrarily, like a garment" (PTG 7).[20]

CHAPTER 2

Opening Nietzsche's Genealogy to "Feminine" Body

A Story of Dynamic Non-dualism and Relation

Among tellers of stories about Nietzsche's texts, a common story is that the identity of the self and the concept break down in Nietzsche's writings. Yet few versions of this story have focused on Nietzsche's view of the formation of the self[1] in the "Second Essay" from *On the Genealogy of Morals* to tell of a dynamic non-dualism[2] that can be said to emerge. In the "Second Essay" (1887/1989), Nietzsche traces a transformation of the subject[3] from its early character, which I will call the "legal subject" (*Rechtssubjekt*)[4] to a subsequent formulation that can be referred to as the unified subject.[5] My reading of the "Second Essay" shows that, for Nietzsche, human beings produce a concept of what it means to be a human being at any given time in Western history according to a reciprocal shaping that occurs among the concept of the subject and the developing constitutions[6] of conscience and punishment. I choose these three constitutions—subjectivity, conscience, and punishment—for two reasons. First, they are prominent themes in *On the Genealogy of Morals*. Second, they include the extremes and center of a continuum that I call Nietzsche's dynamic non-dualism.[7] My rendering of Nietzsche's story provisionally categorizes subjectivity as ideational (the immaterial extreme) conscience as psychosomatic (the center) and corporeal punishment as socio-physical (the material extreme). The borders of these categories are fluid and overlapping.

In this chapter, I show that the emergence of the supposed unified subject coincides for Nietzsche, with an understanding of subjectivity that denies a dynamic non-dualism constituting the subject's formation. In so doing my view of Nietzsche's implicitly enters a current debate among Continental feminist philosophers. This debate begins to open Western philosophy "to something other than traditional Enlightenment rationality" (Oliver 1995, xii). For the purposes of this chapter, *other* means "body"—especially feminine body. Body can be viewed as other because Enlightenment philosophers tend to adopt the realm of immaterial mind as a first principle, and the realm of material body as appendage.

In *Imaginary Bodies: Ethics, Power and Corporeality*, Moira Gatens appeals to a Spinozist monism to introduce "a good deal of dynamism into the categories 'sex' and 'gender'" (Gatens 1996, 149).[8] Indeed, Nietzsche's non-dualism might also be useful for developing a conception of feminine body. Kelly Oliver suggests, however, that although Nietzsche begins to open philosophy to the other, the body, he opens it to only a masculine body (Oliver 1995, 17–25).[9] "It doesn't seem that Nietzsche imagines that wisdom's beloved—this warrior who writes and reads with blood—could be Athena" (24).

Some philosophers, such as Elizabeth Grosz, have proposed ways to continue opening Western Continental philosophy to an other and especially a feminine other. According to Grosz, Nietzsche, Foucault, Freud, and Lacan assume the corporeality of knowledge production, but the "corporeality invoked is itself not concrete or tangible, but ironically, 'philosophical'" (Grosz 1996, 37–38). Traditionally, men have adopted the realm of mind for themselves, writes Grosz. By retreating to the spiritual and "philosophical" to explain the corporeal, Nietzsche embalms the bodily, and eliminates the feminine other who, symbolically, is *the* body (38).

Unfortunately, Grosz too quickly glosses over the meaning of concrete corporeality, especially in the context of Nietzsche, whose non-dualism makes problematic the identities of abstract and concrete. This problem of identification recalls Michael Ondaatje's *The English Patient*: "In Asian gardens you could look at rock and imagine water, you could gaze at a still pool and believe it had the hardness of rock" (Ondaatje 1992, 170). Ondaatje intimates the trickiness of identities like stone and water. Nietzsche's dynamic non-dualism also registers the nuanced deceptiveness of identities, virtually dissolving Cartesian dualism albeit non-reductively. The mental and corporeal become not substantially different in this regard, but neither is one reducible to the other.

So perhaps the complaint should not be that Nietzsche's corporeality lacks tangibility, as Grosz writes, but that his simultaneously tangible and intangible corporeality is as Oliver suggests. It is laden with symbolism traditionally linked with masculine body—the blood of a warrior and not a menstruator, the instruments of an administrator of torture and not those of, for instance, a domestic or a cook. Nietzsche's philosophy of embodiment seems less guilty of disembodying body than excluding what might be called a symbolically feminine body.

Yet menstrual blood and the arts of domesticity and cooking do not apply to all women, nor even, at least in the case of cooking, exclusively to women. Indeed, postmenopausal women do not menstruate, not all women keep house or cook, and many men do. Moreover, housekeeping and cooking are hardly symbols many Western women will want to identify with in any primary way.

In this chapter, by a *symbolically feminine body*, I do not mean a symbolism around which women might rally as women look toward a future. I am not hoping to find an ideal metaphor for a symbolically feminine body. Indeed, the legitimacy of such an ideal metaphor could scarcely be maintained in the context of Nietzsche's critique of moral values. It seems that no value, whether positive or negative attaches absolutely to an item, for Nietzsche. The value of any item appears instead to emerge, transform or dissolve according to the item's context and the evaluator's perspective. By *symbolically feminine body*, then, I mean those images and tropes (irrespective of the value we may attach to them now), which have traditionally shaped, in part, women's value and status in modern Anglo-European culture.

My use of the term *symbolically feminine body* is indebted to Moira Gatens's *imaginary body*. Gatens uses the term *imaginary* in a "loose but nevertheless technical sense to refer to those images, symbols, metaphors and representations which help construct various forms of subjectivity. In this sense, [she is] concerned with the (often unconscious) imaginaries of a specific culture: those ready-made images and symbols through which we make sense of social bodies and which determine, in part, their value, their status and what will be deemed their appropriate treatment" (Gatens 1996, viii).

Importantly, these images and tropes as a symbolically feminine body are not static or unchanging, but are transmogrifying and so, are provisional. On this point, my view of Nietzsche's reflects one of Clifford Geertz's views on the subject of what passes for common sense. The way symbols accrete significance appears to be similar to the way, according to Geertz, a community's collective viewpoint of common sense takes shape. People sharing a communal history may roughly agree on the meanings of certain symbols; and they may roughly agree on whether a particular viewpoint qualifies as common sense. As several cultural anthropologists have indicated, nonetheless, although common sense exists, universal common sense does not (Geertz 1983, 75).

So too, although a symbolically feminine body appears to exist, a universal significance of symbolically feminine body appears not to. Cooking therefore, (or other such tropes—menstruation, weaving, etc.) is not, in this project, intended as a performative ideal toward which women should aspire nor as a static form with which women are stuck, but as an historical image that has dominantly been and in many ways still remains (if provisionally) associated with women.[10]

Curry involves cooking. In my view of Nietzsche's story, curry works on several levels. It operates as a metaphor that contributes to a symbolically feminine body *and* as a metaphor for symbolically feminine body per se. More important, the curry metaphor illustrates Nietzsche's dynamic non-dualism. It displays the field of forces of which the symbolically feminine body is constituted and in which it is participant and factor. Symbols like *curry*, contributing to symbolically feminine body, are co-participants in (and co-constituted by) the immediate and teeming field I call Nietzsche's dynamic non-dualism.

In what follows, with the curry metaphor, I open Nietzsche's discourse of body to a symbolically feminine body that is both possible and questionable. The curry metaphor works on two levels: as a symbolically feminine body and as Nietzsche's conception of subject-formation as a dynamic non-dualism in which the corporeal and the mental share fields of interconnecting forces and affects.

A symbolically feminine body is not only possible but also questionable, because the symbols constituting it do not do so universally or properly. Symbols amass their significance from culturally bound meanings and practices. Thus, a symbolically feminine body that emerges in one set of cultural practices may appear questionable from certain other sets. Like a moving target, a symbolically feminine body often eludes one's intellectual grasp if one attributes to it over time the same imaginary aspects. My story of Nietzsche's non-dualism indicates movement in constitutions like symbolically feminine body and categories like ideational, psychosomatic, or socio-physical. It tells a story of the immaterial (ideational) and material (socio-physical) not as separate realms but as cavernous, overlapping planes.

PART I. RELATION, BREEDING, AND PARADOX

Before I turn to the curry metaphor, let me first expose Nietzsche's view of the process of subject-formation in the West as he forecasts it (perhaps unwittingly) in the opening line of the "Second Essay" (1887/1989). Whether a knowing gesture or a case of Nietzsche saying more than he means, the opening line can be said to anticipate Nietzsche's story of subject-formation. Elizabeth Grosz speaks of texts that "exceed themselves, where they say more than they mean, opening themselves up to a feminine (re)appropriation" (Grosz 1996, 38). In a role of storyteller myself, I too unveil certain submerged themes—themes lurking in shadows and at edges of Nietzsche's story.

My view in this chapter begins as an etymological one. It begins by disclosing an etymological tale of three German terms: *Verwandtes*, *Schuld*, and *heranzüchten* in order to show how Nietzsche's opening line anticipates his story of subject-formation. These etymological details parallel and enhance Nietzsche's story of the subject that Nietzsche displays in the body of his "Second Essay." They suggest that for Nietzsche, the ideas of virtue and God are bred into human beings by discipline and cruelty. More important, they give shape to Nietzsche's view of subject-formation and dynamic non-dualism and my reasons for naming both *relation*.[11]

The name *relation* comes from the title of Nietzsche's "Second Essay" "Guilt, Bad Conscience and the Like." *The Like* is Walter Kaufmann's translation for the German *Verwandtes*, which can also be translated "related things" or "relations."[12] Indeed, Nietzsche tells us not about logically timeless relations among ideas, but about undermining the idea of such an eternal logic. This he does by disclosing a history of relations (genealogy)—one might even say of descendance. He discloses a history that gives rise to historically conditioned Enlightenment descendants: guilt and bad conscience.

Nietzsche considers descendance in terms of human familial relations.[13] Just as the English "relation," the German *Verwandte(r)* can signify a particular human being associated with another by law or by birth. *Verwandte* and *Verwandter* signify the feminine and masculine case nouns, and mean "human relation" or "relative"—that is, a parent, a cousin, and so forth. Genetic familial bonds are relations because the genes among certain people have mixed. An offspring never replicates one or the other of its progenitors but is constituted by a new relation of genes. The notion of genetic relation is significant because it shows that relation, although not exclusively physiological, includes the physiological, the other of Western philosophy.

In addition to familial relations, *Verwandtes* points to other meanings. It is related to several German words: *verwenden* (to apply, to put to use, to give purpose to); *Verwandlung* (metamorphosis, transformation; *verwandeln* (to transfigure, to transform); and *wenden* (to turn, to exchange). The signifier for the adjective *verwandt* is the same as that for the simple past tense of the verb *verwenden*, which means "to use." One can understand *verwenden* as that practice by which a human being gives a purpose to a constitution. To put a constitution to use, one must propose a use (or meaning) for that item. Once so meant the constitution becomes related (*verwandt*) to that assigned purpose. It becomes a vehicle for achieving a desired goal or signifying a desired signification. Thus an item's use appears to be a matter of convention and not of some immutable nature.

Interestingly, an additional relative of *Verwandtes* (and the title of Franz Kafka's well-known short story) is *die Verwandlung*, meaning "metamorphosis" or "transformation." *Verwandlung* comes from the verb *verwandeln*, which means "to transform." Moreover, if one takes a closer look at *verwenden*, one notices its stem: *wenden*. *Wenden* means "to turn." It can signify a turning about or round (a top or a wheel), a turning towards (a ship turning toward harbor), or a turning in the sense of an exchange (money for property). Thus, *Verwandtes* (relation) suggests a descendant infused with an historically conditioned meaning or purpose. It can undergo transformation, turn towards a new end, or exact an exchange.[14]

My etymological story draws attention to this kind of transformation in the case of two relatives: *Schuld* (guilt) and *Schulden* (debt). In the relation between these, Nietzsche discloses not only a change in *meaning* and *purpose* regarding the root *Schuld* but also a significant change in its kind of constitution. The "major moral concept *Schuld* [guilt] has its origin in the material concept *Schulden* [debts]" (GM 2, 4). Nietzsche tells of an early custom informing the material (socio-physical) constitution *Schulden.* This intimates a transformative history from which the later psychosomatic[15] constitution *Schuld* (guilt) emerges. Important here is that *das Verwenden* (putting to use) and *das Wenden* (turning) can occur in such a way that not only a change in the purpose and meaning of a custom occurs, but a *Verwandlung* (metamorphosis) of a constitution kind (socio-physical, psychosomatic, ideational) may take place. In the case of *Schulden*—a socio-physical constitution gives rise to a psychosomatic constitution.

I have discussed genealogical relation (*Verwandte*) and its link to "putting to use" (*das Verwenden*), turning and exchange (*das Wenden*), transforming (*das Verwandeln*), and metamorphosis (*Verwandlung*). *Verwandlung* brings forth the idea of breeding in the context of human formation. Nietzsche implies that breeding, in the case of humans, involves transformation (*Verwandlung)* and putting to use (*verwenden*). The first line of the essay reads, "To breed [*heranzüchten*] an animal *with the ability to make promises*—is not this the paradoxical [*paradoxe*] task that nature has set itself in the case of man" (GM 2, 1)?[16]

"Nature" qua human animal, Nietzsche's line suggests, cultivates the capacity to remember and to reason in itself. To give rise to an animal that can make promises, one must first "*make* [Nietzsche's emphasis]" (*zu machen*) a human animal "to a certain degree necessary, uniform, like among like, regular and consequently calculable" (GM 2, 2). "The tremendous labor of that which I have called 'morality of mores' (*Dawn*, sections 9, 14, 16)—the labor performed by man upon himself during the greater part of the existence of the human race, his entire *prehistoric* labor, finds in this [the making of a calculable human] its meaning, its great justification" (GM 2, 2). These lines of Nietzsche's story indicate that the breeding of a "uniform" and "regular" human out of human that is neither can be viewed as exemplary of the practice of "putting to use." Amidst changing circumstances and mortal desires, human animal shapes and informs itself with temporary meanings to suit its will. The predominate, albeit temporary, meaning guiding human being's "prehistoric labor" concerns turning oneself into an animal that can make promises. In this sense, one might say "prehistoric" human being interprets itself and puts itself to use (*verwenden*) with this aim in mind. It is significant, then, that this gradual breeding process excludes a transcendent will or essential design. It suggests instead a dynamic process of co-constituting forces within and across the permeable borders of the subject.

Nietzsche's story in his "Second Essay" suggests that human animals participate in a process that continually and subtly breeds (*heranzüchten*) human selves in new forms. These transformations occur as a gradual result of certain necessary factors, including our own particular interests and wills.[17]

Moreover, Nietzsche's vision describes this breeding (*heranzüchten* or *züchten*[18]) of an animal that promises as a "paradoxical task" (GM 2, 1). This description brings forth an idea that I call Nietzsche's idea of paradox. At least one paradox regarding the formation of human subjectivity is implicit in Nietzsche's story as he tells it in his "Second Essay." Human memory, reason, and the idea of God arise gradually out of their own absence. The process according to which Nietzsche describes this emergence indicates in nature an unmeasured will rather than a preordained design.

Nietzsche's paradoxical tale suggests that his origin not only involves "reason," "virtue," and "God," but also discipline, blood, and imprisonment. If one reflects on the term *heranzüchten* in the essay's first line, one can ascribe to the opening of the essay a link not only to the concept of virtue but also to the concept and practices of punishment. Although the nominative for *züchten*–*Zücht*–may mean "the act or process of cultivating" (plants, animals, etc.), *Zücht* is more commonly used to signify "the practice of discipline." One disciplines one's children and a state punishes its criminals. There is also the verb *Züchtigen*, which means "to beat or to flog–that is, to employ corporeal punishment." Related to this is *Züchthaus*–"a place to send those in need of discipline or punishment." *Züchthaus* is the German word for prison. *Züchtig*, alternatively, bears a different meaning from the others. One attributes this adjective to "one who shows modesty, innocence or a sense of virtue." A chaste or unassuming young boy or girl, or a responsible or "moral" adult, is said to be *züchtig*. In *heranzüchten*, then, Nietzsche anticipates his ensuing tale in his "Second Essay," a tale of the origins of humans in the West. In this story, the ideas of virtue and God are bred into human beings by discipline and cruelty. Of interest is not only that virtue and God are *Verwandte* (relatives) of corporeal punishment, but also the implicit intimacy of that relatedness. The chaste, "moral" adult has become that way especially, Nietzsche's story indicates, due to a personal history of discipline and punishment. From this one can gather that a perceived modern intimacy between punishment and purity may be the latest incarnation of a protracted lineage of correction and physical cruelty (GM 2, 2).

Nietzsche's view of the breeding of the human animal is significant because it undermines the Enlightenment notion of human nature, reason, and God as preeminently timeless. I have shown this by developing the concept of relation as *Verwandte* in each of its senses: *Verwandte, verwenden, Verwandlung,* and *wenden.* For Nietzsche, the concept of relation shows that a constitution categorized as the self and virtually all other so-called "kinds" of constitution as well, exist as that "kind" only provisionally. The concept of relation illustrates that *what* a constitution is, that is, a constitution's category or kind is not permanent or universally real.

According to the notion of relation, what a constitution is is provisional because of *how* constitutions exist. Earlier, I categorized three kinds of constitutions: ideational, psychosomatic, and socio-physical. Yet the kind of each is inextricable from the *how* of each and that how is a relatedness, *Verwandte,* even more, a relatedness among "kinds." An etymological argument about *Verwandte* is a way of showing that to which the concept of *Verwandte* (relation) points. Its etymology discloses what it signifies—that the so-called "ideational," "psychosomatic," and "socio-physical" are co-extensive. We can begin to see such co-extension in the examples socio-physical *Schulden* (debt) gradually becoming psychosomatic *Schuld* (guilt) and socio-physical punishment giving rise to moral purity.

Nietzsche's story of the subject is a story about a process of fluidity and formation—transformation. More important, the concept of *Verwandte* (relation) as transformation undermines a Cartesian dualism and points to a concept and experience of dynamic non-dualism. If one discerns experience as dynamically non-dual, then the sort of analysis presupposing an essential separation between life and values used by Edward Andrew to point to the costs of Nietzsche's value-discourse, loses its basis (Andrew 1999, 72).[19] According to a dynamic non-dualism, the provisional categories of ideational, psychosomatic, and socio-physical are revealed as provisional precisely because their hazy borders overlap and appear in the end, not to be in any strict sense, borders at all. The forces of the ideational, the psychosomatic, and the socio-physical are related then, inextricably, as if along a dynamic continuum. For this reason, the concept of relation constitutes a concept and experience of dynamic non-dualism.

PART II. THE FORMATION AND RELATION OF CURRY

I have roughly defined the concept of relation, shown that subject-formation as relation puts into question a transcendent notion of God and human nature, and shown that relation can be said to constitute a dynamic non-dualism. In my analysis of Nietzsche's, I have also shown that the opening line of Nietzsche's "Second Essay" implicitly forecasts Nietzsche's tale of human animal's lineage—a lineage that unfolds throughout his essay. Now my examination turns to the more explicit relation forming the subject in Nietzsche's "Second Essay." Because the relation forming the subject is complex, I begin by bringing relation into relief through the curry metaphor. Nietzsche never mentions curry. I open Nietzsche's discourse to the metaphor of curry (and implicitly of cooking) to create a framework through which the possibility of a symbolically feminine body can emerge. Curry operates not only as a metaphor for Nietzsche's dynamic non-dualism but also as metaphor for a symbolically feminine body—a body that often physiologically and symbolically has been bound not to the "philosophical" but to the "material," that is, earth, cyclical blood, and home cooking.[20]

Nietzsche's dynamic relation can be shown to constitute a cooking flavor known in the West as curry. *Khardi*, the name of an Indian yogurt dish, gave the English language the word "curry"—a term that in the minds of many of us in Western societies describes Indian food. The dish gets its name from its dominant spice—a stalk of fresh *Khardi* leaves. Ironically, the name of the leaf has no special relation to the flavor many Westerners understand as curry. In fact, *khardi* is not among the ingredients of U.S. store-bought curry, although it arises in some spicy and hot Indian recipes.[21]

For many of us in the West, "curry" signifies a relatively constant and fixed cooking ingredient. Indeed, unless one makes one's own curry, the mass-marketed version is rather uniform. For instance, McCormick's brand of curry powder—similar to other brands—combines fenugreek, coriander, cumin, turmeric, celery seed, mace, ginger, red and black pepper, and garlic.

Another conception of curry is that of curry as relation. Each of the nine curry recipes I consult appears in Sumana Ray's *Indian Vegetarian Cooking* (Ray 1990). Ray's book, originally published in England for a Western public, integrates the term "curry" into a number of her English translations of Indian dish titles (few of which includes the word *Khardi*). Ray's method of cooking reinforces the idea of *relation*. It reinforces the idea that spices and food, like the constitutions of punishment, conscience, and subjectivity are matters of reciprocal shaping and interpretation—a fluid *relation*.

When Ray makes curry she weighs the respective tastes or meanings of each ingredient (that is, of the spices and the items to be cooked), in terms of the relation she anticipates will transpire. The "relation of curry" operates like a double genitive.[22] It reveals each ingredient in a simultaneous role as agent and patient. It reveals the identity of each ingredient to be permeable to inner and outer forces—undergoing change among its co-ingredients.

Depending upon the texture, existing taste, and flavor of what she is cooking, Ray includes or excludes certain Indian spices often associated with curry. Interestingly, there is no single spice that she includes in every one of her curries. Curry, when considered in many Indian culinary traditions, shows itself to be a concept that simplifies its complex referent—an elusive constitution of multiple ingredients in which no ingredient remains constant.

In contrast to understanding curry as a fluctuating spice-relation, is a predominant Western idea of it. As the supermarket brand of McCormick curry exemplifies, the U.S. notion of curry is steeped in a tradition of simplification and uniformity. In the essay "Hygiene and Repression," Octavio Paz perceives American cooking as a translation of a puritanical lineage bent on separation and exclusion (Paz 1994). The peculiar American heritage combining innocence before God and faith in the perfect truths of scientific reason, spills into our cuisine: "Traditional American cooking is a cuisine without mystery: simple, nourishing, scantily seasoned foods. No tricks: a carrot is a homely, honest carrot, a potato is not ashamed of its humble condition, and a steak is a big, bloody hunk of meat" (16). Arguably for many of us in the United States, curry is the supermarket version—simple, fixed, and definable: a greenish element that might transform most any dish into Southeast-Asian fare. For many of us in the United States, curry has become—or has always been perceived as—an unchanging form.

Curry in this sense can be said to exemplify practices involving rough-hewn generalizing rather than finely tuned deliberating. Supermarket curry includes more spices than many singular Southeast Asian curries; this may be to make curry attract a wider range of taste buds and to make it applicable to a broader range of dishes. For Nietzsche, a cook (or philosopher) who abandons his or her opportunity to choose the appropriate spices for an occasion might be said to have "bad" taste. "Verily, I also do not like those who consider everything good and this world the best. Such men I call the omni-satisfied. Omni-satisfaction, which knows how to taste everything, that is not the best taste. I honor the recalcitrant choosy tongues and stomachs, which have learned to say 'I' and 'yes' and 'no.' But to chew and digest everything—that is truly the swine's manner" (Z 3, 11, 2).

PART III. THE FORMATION AND RELATION OF THE SUBJECT

Just as in Ray's Indian cooking different spices are combined each time one prepares a food "to taste," in Nietzsche's tales of human origins, different force combinations are shown to constitute the meaning of the subject ("ideational"), conscience ("psychosomatic"), and corporeal punishment ("socio-physical") from one moment in history to the next. Here I place the former categories in quotes to indicate their categorial provisionality. This provisionality can be understood in light of my account of relation, which like curry points to the converging borders of constitutions. Each constitution appears to represent a changing relation. Moreover, as cumin, turmeric, and mustard seed simmered together reciprocally shape one another, so do the constitutions of the subject, conscience, and corporeal punishment. What it means to be a human being at any given time in Western history appears to arise according to a mutual shaping that occurs amidst the blending fields of the subject and the developing constitutions of conscience and corporeal punishment. The character of the subject seems like that of Indian curry. It seems to possess no constant ingredient. Alongside conscience and corporeal punishment, the subject continues to change like the intensified commingling of spices in day-old curried soup. Because of this continuous emitting and receiving of meaning by the subject, I call the definition of the subject a relation.

Early in Nietzsche's "Second Essay" the breeding of an animal with the ability to make promises, that is, the breeding of a supposed unified subject, presupposes the "making" of memory (GM 2, 2). According to Nietzsche's *On the Genealogy of Morals*, the early mixture of human formation has ingredients for this: a legal subject (i.e., a disunified subject), an animal with mind, and a penal custom. However, these need to simmer. One can taste Nietzsche's genealogical mixture along the way until it yields a combination so cold it burns and leaves an imprint of a Christian God on a numbed palate. This latter development, suggests Nietzsche, is stirred by a subject that views itself as a moral unity, a conscience that does not clear itself,[23] and a penal custom that has become internal (guilt).

Arguably for Nietzsche, the early meaning or the "taste" of the legal subject (*Rechtssubjekte*[24]) is influenced by its two complementary ingredients: forgetful mind[25] and corporeal punishment. This kind of subject is as old as "the contractual relationship between *creditor* and *debtor*," which "points back to the fundamental forms of buying, selling, barter, trade, and traffic" (GM 2, 4). It was in the contractual relationship "that *promises* were made, and that *memory* had to be *made* for those who promised" (GM 2, 5). Memory, Nietzsche's tale says, has not yet been "made," but a process of its formation is underway. The legal subject possesses an unformed memory. Unlike the formed memory/conscience that will succeed it, unformed memory still has the ability to clear itself. It can forget. Therefore, it seems that for Nietzsche, the definition of the legal subject includes an aspect of its associate: the forgetting mind. Like a cardamom seed simmering with cloves, the legal subject "means" or "tastes" more like legal subject than forgetting mind. Nevertheless, just as the flavor of the cardamom seed could not be described without reference to its modifying cloves, the meaning of legal subject cannot be described without reference to its modifier "forgetting mind."

Nietzsche's genealogy suggests that the legal subject also has another partner: corporeal punishment. When a legal subject makes a contract with a creditor, the contract promises the creditor that some form of just compensation will be paid even if the debtor cannot repay with money (GM 2, 5). This pledges that should the debtor have no money, then he would offer "something else that he 'possessed,'" as compensation—"his body, his wife, his freedom, or even his life" (GM 2, 5). In the era of the legal subject, the penal custom ensures the creditor that he or she will receive some form of return. Thus, both forgetting mind and corporeal punishment inform the definition[26] of legal subject.

The reciprocity among constitutions broadly depicts *relation* (as *Verwandte*, *verwandeln*, *Verwandlung*, *verwenden*, and *wenden*) and constitutes Nietzsche's dynamic non-dualism. The forces of the legal subject, forgetting mind, and corporeal punishment interrelate across converging planes. They separate, shift, and convene across an apparent continuum. The dynamism of relation can also be described more specifically. Two of the significations for *wenden* are to turn towards and to exchange. Characteristic of Nietzsche's legal subject is the habit of turning an item (*wenden*) towards a particular use (*verwenden*). One invests it with a meaning that relates it to other items bearing a similar purpose or meaning. This is what happens in the case of a creditor who agrees that physical pain leveled upon his or her debtor is an equivalent to the money that the debtor cannot repay. Corporeal punishment is turned towards the end, that is, it procures the meaning of money. Thus, not only is there an exchange (*wenden*) of corporeal cruelty for money but the utility of pain becomes transformed (*verwandelt*). It becomes a substitute for money and so adopts the usefulness of money, marking another kind of relation, that of *verwenden*, to apply or put to use.

The possibility of such a turning to a particular use proves a distinguishing quality of the legal subject. For unlike the subsequent supposed unified subject, the legal subject participates in an era in which debts are consistently discharged. The debtor either pays with money or with something deemed *exchangeable* (*das Wenden*) for money, in other words, excruciating pain. "The oldest and naivest moral canon of *justice*" prevails: "everything has its price [*Preis*]; *all* things can be paid for [*abgezahlt werden*]" (GM 2, 8). The debtor pays the creditor back by giving the creditor "a kind of *pleasure*—the pleasure of being allowed to vent his power freely upon one who is powerless" (GM 2, 5). Released of his or her debt, the tortured debtor is free to continue his or her life—conscience cleared. Conscience in the era of Nietzsche's legal subject has not yet become what Nietzsche calls "bad conscience."

I have generally characterized Nietzsche's legal subject, and have shown the forces that will eventually transform a mind that forgets into a mind that remembers. Nietzsche implies that only a mind that remembers—that is, with memory—could originate ideal forms and immanent essences. Memory is a facet contributing to the formation of bad conscience and the supposed unified subject. However, it does not account for another facet, that which makes our bodily instincts turn inward, back upon ourselves. Nietzsche gives two reasons for this, one of which I discuss.[27]

In *On the Genealogy of Morals*, Nietzsche hypothesizes about the emergence of the idea of God in the context of the tribal community. Out of a particular mix of relations the idea of God arises. The size of one's debt and the extent of one's fear of the ancestor increases proportionately with the expansion of a tribe's power (GM 2, 19). If one carries this psychic punishment to its extreme, the people of the "*most powerful* tribes" will project their ancestors to "monstrous dimensions through the imagination of growing fear" (GM 2, 19) and of debt. Eventually, the seemingly immutable quality of one's self-afflicted psychic punishment reflects an ancestor of similar quality, one that is fixed and eternal.

The more powerful, controlling, and organized a community grows, the more divine become its ancestors. The Christian God represents the apex of such order and control. It also signifies the largest accumulation of guilty indebtedness.

The emergence of the so-called unified subject coincides with such a development of internalized guilt. Although the formation of the concept of a Christian God is said to come from a movement that I describe as a relation (in the sense of familial relative, putting to use, turning towards, exchange, and transformation), its arrival signals in Nietzsche's story, paradoxically, a freeze cold enough to burn and an ideal that captures this burning immobility: God. Instead of allowing the change and mobility marked by relation to continue, human beings—Nietzsche's story implies—will the movement to stop.

In this "moralization" process, Nietzsche's tale tells us, human beings reveal their preference to halt once and for all the movement by which debt and guilt can be paid for. Via a mixture of two constitutions in particular: "*bad* conscience [and] the concept of god," the Westerner's guilt appears insoluble. "The *aim* now is to preclude pessimistically, once and for all, the prospect of a final discharge"(GM 2, 21). One wills the possibility of turning (*wenden*) one's payment into a payment of cruelty, the possibility of there being this or other items that can be transformed (*verwandelt*) into the equivalent of one's debt, to be lost. The process by which the legal subject puts an item (such as corporeal punishment) to a particular use (*verwenden*) to mean the equivalency of something else that one owes another (an amount of money), has been choked off. "At last the irredeemable debt gives rise to the conception of irredeemable penance" (GM 2, 21). For Nietzsche, the idea of God leaves the conscience frozen stiff and therefore "unified."

Just as supermarket brands of curry deny a factor so crucial to many Southeast Asian curries—the factor of relation—faith in the unified subject covers over a view of history and life as relation without *relata*.[28] Like the moralist tradition that Nietzsche's tale of origins shakes up, the idea of a supposed morally integrated, unified subject presupposes a permanent link between origin and purpose. For Nietzsche, it seems, this view is not merely wrong but worse: it is in bad taste!

Moreover, like packaged curry and the submersion of desire in the American culinary tradition, the Western ideal of the unified subject, Nietzsche argues, seems to suppress life. It seems to curtail one's ability to expand, to move, to overcome the self. It intimates, as Nietzsche's playful intestinal metaphors[29] imply, that ill health might best be viewed as a symptom of poor digestion.

PART IV. TO *CURRY* A BRIDGE

Recall in the first line of Nietzsche's "Second Essay" that Nietzsche begins by suggesting the unfurling of a paradox: "to breed [*heranzüchten*] an animal *with the ability to promise* . . . is not this the paradoxical task that nature has set itself" (GM 2, 1). Nietzsche traces a transfiguring *relation* that extends from the disintegrated legal subject to the fateful rise of the apparently unified subject. Its genealogy of dynamism and permutation shows Nietzsche's tale of origins undermining a mind-body dualism and promoting a dynamic non-dualism. Indeed, the more or less ideational, psychosomatic, or socio-physical constitutions at play in my story of his are categorized as such only provisionally. Constitutions appear to be interlaced and influencing each other. Virtually none seems identifiable as solely or permanently ideational, psychosomatic, or socio-physical, nor as solely or permanently a singular item per se. The dynamism points to the unlikelihood of such permanence and identifiability.

Against this backdrop, Nietzsche would have us reconsider our subjectivity. That an idea such as a unified subject uniting origin and purpose emerges from a formative mix of factors in which origin and purpose are separate and changing, is paradoxical, suggests Nietzsche. Even more uncanny, Nietzsche implies, is that "reason," "virtue," and "God" appear to be the descendants of discipline, blood, and imprisonment.

My analysis explores this uncanniness by looking at Nietzsche's treatment of subject-formation in his "Second Essay." It finds that for Nietzsche, the legal subject, forgetting mind and corporeal punishment appear to be a relation; moreover, that relation constitutes a dynamic non-dualism. My development of the curry metaphor illuminates Nietzsche's non-dualism as relation, and operates as a metaphor for Nietzsche's non-dualism and/as relation. The dynamism of Nietzsche's relational subject, like the dynamism of Indian curry, shows that the "kind" of each constitution (i.e., legal subject or unified subject; forgetting mind or conscience) is inextricable from the *how* of each. That *how* appears to be a relatedness, *Verwandte*, indeed, a relatedness among kinds of items.

The three most broadly conceived kinds of constitutions—ideational, psychosomatic, and socio-physical—appear relating like cavernous overlapping planes. Such relating points to why Nietzsche's story can be said to put traditional tales of Cartesian dualism into question. The mental (ideational) appears to be not purely so, and likewise the material (socio-physical). Entities which my view of Nietzsche's story categorizes "ideational" (for instance, a symbolically feminine body and subjectivity) are not ideational in a permanent or universal way. This is due to *how* entities appear to exist. Their borders overlap, remaining permeable to inner and outer influences.

Specifically, I have focused on an exacted metamorphosis of one kind of entity into another. I have focused on Nietzsche's trace of the transformation of the legal subject into the so-called unified subject, that is, the transformation of the legal subject into a supposed human nature that is coherent and rational at its fundament and core. If the unified subject appears to be for Nietzsche a co-construction in a sea of myriad forces, including co-participants conscience and corporeal punishment, then the so-called unified subject seems to be not unified at all. It appears to possess no reified or universally real stature. My view of Nietzsche's non-dualism and/as relation—illustrated in part through the curry metaphor—reveals a field of swarming, cross-pollinating forces constituting these co-constructs.

The causality, according to which the transformation of the legal subject to the supposed unified subject seems to occur, is not the traditional linking of Origin and purpose. A mixture of interlacing forces constitutes this causality, including human preference and irrational fate. The causality appears to be not traditionally Original but relationally originating. It appears to work according to a purposeless necessity[30] in which power relations and blind destiny replace the logic Nietzsche attributes to Socratic and Aristotelian *doxa*. Thus for Nietzsche, his "Second Essay" writes literally *para doxa*.

Nietzsche does not claim to know whether his story of relations is right, but he asserts it as a more probable and aesthetically appealing interpretation of reality than most. If Nietzsche's non-dualism as relation correctly articulates the circumstances making up the legal subject, forgetting mind and corporeal punishment, then the subsequently generated notion of the so-called unified subject is in the end (and from the beginning) misguided.

The curry metaphor for Nietzsche's dynamic non-dualism and/as relation reflects an illusory concept that is the so-called unified subject. More important, the Indian curry metaphor also works to represent symbolically feminine body. It ties together Nietzsche's dynamic non-dualism with symbolically feminine body. Prompted by the logic, if not the symbolism, of Nietzsche's words, this essay builds (or "curries"?) a bridge from Nietzsche's writing to a symbolically feminine body.

Yet as it "curries" this passage, one ought not to forget the tenuousness of that toward which it stretches—a possible symbolically feminine body. As many anthropologists have shown, communal history and culture endow a group's language and habitudes with symbols unique to its heritage. If a symbol such as curry contributes to a symbolically feminine body according to a particular Anglo-American perspective, the symbol remains questionable from certain other perspectives, some of which may originate within contemporary Anglo-European culture. And so, if a symbolically feminine body is possible, it is also already at the start—from at least some perspective—questionable.

Nietzsche's thought, although beginning to introduce the other qua body into Western philosophy, introduces primarily symbolically masculine body. Even so, the being of such a body, constituted by and co-participant in a moving constellation of partnering forces may be, if not feminine, feminist. This is because it indicates that kinds of bodies (that is, masculine, feminine, mental, physical) cross-germinate and gradually metamorphose. They are not ironclad with a form-giving code constraining them like a giant buckle to traditional Western roles.

In the tradition of the West, and especially in accord with on-going patterns of the Enlightenment, feminine bodies have been viewed as entities imbued with an indelible nature proper to women. Such a body can be called "natural" woman. Jean Jacques Rousseau offers such a body as his view of woman in *Emile or On Education* (1762/1979). Woman is naturally "passive and weak"; man "active and strong" (358). From their disparity in physical power "it follows that woman is made specially to please man" (358). Nature spells out woman's natural position as child bearer and caretaker, says Rousseau. "Everything constantly recalls her sex to her; and, to fulfill its functions well, she needs a constitution which corresponds to it. She needs care during her pregnancy; she needs rest at the time of childbirth; she needs a soft and sedentary life to suckle her children; she needs patience and gentleness, a zeal and an affection that nothing can rebuff in order to raise her children" (361). Alongside Rousseau's idea of a weak, child rearing, and disfranchised natural woman stands his natural man: strong, rational, and franchised.[31] Nietzsche's "Second Essay" does not explicitly expose the concept of "natural" woman as a mistake. But by intimating the concept of the unified subject (natural man) as mistake, Nietzsche's critique extends to the concept of natural woman. The possibility of a unified subject depends upon a principle of permanence in human nature, like the Platonic forms. Nietzsche's dynamic non-dualism theorizes against such permanence.

If Nietzsche's thought explicitly unravels the concept of a supposed unified subject, it implicitly unravels the concept of a supposed natural woman. With respect to the idea of natural woman, Nietzsche's words "say more than they mean" (Grosz 1996, 38). Like a bridge (or Zarathustra's tightrope?), they can be shaped to arc toward a feminist tale.

Finally, the tendency among recent feminist philosophers of embodiment has been to avoid a notion of body as biology while maintaining a non-dualist model of mind and body. Gatens makes clear in *Imaginary Bodies* (1996) that by "body" she does not mean biology but symbolic body. Although body as biology *only* is problematic, for the nonreductive non-dualist for whom what it means to be a human being can be reduced neither to the mental nor to the biological, body as symbolic *only* is also problematic. It reduces all to the mental.

My view of Nietzsche's story offers an account of symbolic body that avoids this reduction. It categorizes the metaphors and symbols making up symbolically feminine body as ideational. Yet, according to Nietzsche's dynamic non-dualism, the ideational by definition is not purely "idea." It interfaces with the psychosomatic and socio-physical, that is, with so-called matter. Nietzsche's non-dualism shows why physiology need not be excluded from a conception of feminine body; it attempts to overcome the contradiction inherent in a supposed nonreductive non-dualist conception of body, which nonetheless excludes the physiological.[32] According to Nietzsche's non-dualism, "biology" participates, if not dominantly, in a symbolism of feminine body. This logic of dynamic non-dualism can be extended to a symbolism of masculine body. This makes a symbolically feminine or a symbolically masculine body a more convincing nonreductive non-dualism and still avoids the suggestion that a woman or a man *is* her or his biology in any permanent way or that such biology is static and thereby limits her or him to traditional roles.

CHAPTER 3

Nietzsche's Ascetic Ideals and a Process of the Production of Embodied Meaning

PART I. INTRODUCTION

According to the concept of dynamic non-dualism, our felt and imagined experience appears to be constituted both within and across the planes of the mental and the physiological, the subject and the object. In and across them, a symbolically feminine body shows itself to be constituted not only by ideational forces but also by socio-physical forces—those implied by what has been called "biology," for instance. It stands to reason that not only the phenomenon of a symbolically feminine body, but other phenomena as well could be shown to be co-constituted by factors not traditionally associated with them in the West. Here I show Nietzsche's concept of the ascetic ideal as and through a metaphor for a process of the production of meaning. The process shows phenomena arising and transpiring according to the concepts of dynamic non-dualism and relation. The curry metaphor in chapter 2 illustrates characteristics of dynamic non-dualism and relatedness that remain suppressed and unrecognized amidst predominating Industrio-Enlightenment perspectives in the West. A metaphor emerging in this essay parallels such a covering-over, this time with respect to Nietzsche's discourse on ascetic ideals. For Nietzsche, a practice of denying ways phenomenal experience reflects concepts of dynamic non-dualism rather than concepts of coherence or transcendence, reaches an extreme with the ascetic priest's view of ascetic ideals. For the ascetic priest, ascetic ideals conform to a conceptual unity.

In *On the Genealogy of Morals* the title of Nietzsche's "Third Essay" is, "What is the Meaning of Ascetic Ideals?" Amid much of Nietzsche's ascetic ideal discourse, the concept of the ascetic ideal shows itself to be indeterminate in meaning and thus, with respect to a semantic referent, indicating an absence of identity.[1] The *meaning* of "ascetic ideals" discloses a paradox between plural meaning[2] and unity. The paradox involves, on the one hand, multiple meanings of ascetic ideals and, on the other, a momentary trumping of these meanings—as we shall see—by one among them, that of the ascetic priest.

In this context, the ascetic ideal represents a signifier communicating and being interpreted in terms of many meanings and so having no single transcendent signification. More significant, in *On the Genealogy of Morals* the rise of the ascetic ideal in human experience represents the emergence of apparent meaning for human existence. Thus, the essay's interrogatory title "What is the Meaning of Ascetic Ideals?" can also be worded "What is the Meaning of Meaning?"[3]

With the exception of the paradoxical emergence of the ascetic priest, the concept of "meaning," like the concept of "ascetic ideals," shows itself in Nietzsche's discourse of ascetic ideals, bereft of determinacy. Its indeterminacy displays a relatedness that is reflected in the grammatical structure forming the phrases "the meaning of ascetic ideals" and "the meaning of meaning." This grammatical formation is sometimes called a "double genitive." The grammatical structure of sentences using a double genitive implies that the subject and the object in the genitive phrase are ambiguous. Thus, in the first phrase, "ascetic ideals" can be interpreted as either the subject or the object of "meaning" or both, and in the second phrase the same can be said for "meaning." The arrival of the ascetic priest, however, challenges this relatedness and reveals paradigmatically a dominating force of the genealogist's inheritance: an apparently unified subjectivity and, correspondingly, an apparently unified concept of the ascetic ideal. A discourse of the so-called unified subject, unsurprisingly, privileges the simplified products of consciousness over the variegated modes of sensation and structures of thought rooted in experience that has not yet become dominantly reflective experience.

The question of the meaning of ascetic ideals, which is indeed the focus of this chapter, is especially abstract. It not only asks about the meaning of an abstract *constitution*[4] (i.e., ascetic ideals or meaning), but it also asks the meaning of a particular abstract constitution: the meaning of *meaning* itself. Because of the obvious complexity here, I will begin with a more "concrete" topic to introduce the characteristic conflating of subject and object that occurs in sentences structured by genitive phrases and to facilitate my analysis of the meaning of ascetic ideals for Nietzsche. The preliminary model will address a more specific question about embodied experience. My criteria for choosing the specific question for the preliminary model are threefold: 1) that the topic of the question is a pertinent object of debate in twentieth and twenty-first century discourses about emodiment and thus, would imply how Nietzsche's writings could be said to matter for such debates; 2) that with respect to the three overlapping planes discussed in chapter 2—the ideational, psychosomatic, and socio-physical—the topic is located (provisionally) amidst the planes of either the psychosomatic or the socio-physical. In other words, the primary way that the topic presents itself experientially would not be in the mode of abstract idea; 3) the question is framed as a genitive phrase that begins "What is the meaning of . . . ?"—paralleling the genitive phrase: "What is the meaning of ascetics ideals?" and thus providing additional opportunity for showing the basic structure of relatedness implicit in a genitive phrase and ultimately, for displaying a relatedness characteristic of the concept and production of *meaning* per se.

Before we discuss the preliminary model, it is worthwhile to first discuss the background of embodiment debates in which the model emerges and to which it responds. Many feminist philosophers of recent decades have argued for a view of body that avoids a Cartesian dualism. Some make clear when talking about mind and body as not two separate substances but as somehow intermixed, that by "body" they do not mean biology.[5] For instance, Moira Gatens means by body the "imaginary body," that is, forms of subjectivity constructed by images, tropes, and symbols socially linked to gendered persons (Gatens 1996, viii). For many feminist philosophers of embodiment, to view body as biology—if even just in part so—is to embrace body as something static, reducible, and nonnegotiable. Elizabeth Grosz, on the other hand, criticizes feminist and cultural theorists like Gatens "who insist on the *discursivisation* . . .of bodies as a mode of protecting themselves from their materiality. Analyses of the *representation* of bodies abound, but bodies in their material variety still wait to be thought" (Grosz 1995, 31). Even so, she, like the theorists she criticizes, is careful to exclude a concept of biology from her non-dualist notion of the so-called material body. "Nonbiologistic, nonreductive accounts of the body" are those accounts best suited "to reposition women's relations to the production of knowledges," writes Grosz (1995, 31). Again, we see that Grosz, like many other feminist theorists, considers a dominantly biological account of body inadequate.

A biological account of body has been viewed as reductive, and theorists of biological body have been perceived as attempting to achieve so-called objective truth about bodies. As Helen Longino shows, such a theorist presumes to exclude his or her own value judgments and political commitments (1990, 220–232). Moreover, he or she generally implies that the results of one's methodology are generally free of socio-historical influence. Such presumptions and their scientific results have understandably precipitated much feminist criticism.

Female biology has often been presented as evidence of nature's supposed fixed role for women. Women are meant to mother and nurture, an outmoded story says. And so, some feminist philosophers of embodiment, although nondualists, conceive of a body that excludes bodies in their material variety.

The view of embodiment offered in this essay likewise avoids a Cartesian dualism. My view of body, however, will differ from some contemporary feminist articulations. My concept of body includes a concept of biology. By "biology," I mean statistics and concepts about life *as* socio-physical. Such information relies on collections of empirical data. At the same time, I believe, like many of our era, that the scientific method is historically and culturally contextual, that the knowledge it produces is qualified, not absolute, and that such knowledge is not necessarily more reliable or of greater epistemological worth than the knowledge a certain nonscientific practice or religion may produce.

In this sense, my view converges with many contemporary anthropologists, such as Thomas Buckley and Alma Gottlieb, who temper their use of scientific methods in collecting and analyzing data, while guarding against the tendency "to view our scientific biology as independent of historical context" (1988, 42). Anthropologists "presume . . . to 'know' what [an entity] really is and turn our attention to non-scientific 'belief' systems armed with this knowledge" (Buckley & Gottlieb 1988, 43).

The scientist, anthropologist, and philosopher are accustomed to viewing nonscientific or religious knowledge as a "smokescreen created to ensure the adherence of simple peoples to procedures that guarantee their physical survival: procedures that, presumably these peoples—seen by turn as acutely observant—would otherwise ignore" (Buckley & Gottlieb 1988, 23). This critical anthropological perspective is probably right, from a certain Nietzschean perspective. Confidence in scientific truth can be said to display a certain naiveté. One is reminded of Nietzsche's well-known comparison in *On the Genealogy of Morals* (3, 24–25) of the scientist's "faith in truth" as the "latest and noblest form" of ascetic priestly and religious faith. It too can be seen as a "smokescreen created to ensure the adherence of simple peoples to procedures that guarantee their physical survival" (Buckley & Gottlieb, 1989, 23). If scientific biology—like religious knowledge—is like a smokescreen, it remains an important one to view; it shows itself to be a persisting factor in the production of meaning in the West.

PART II. THE MEANING OF (PRE)MENSES

One can introduce the process of the production of meaning in the West—shown in Nietzsche's essay on ascetic ideals—with a preliminary model that will also serve as a metaphor. The metaphor I will use for the concept of ascetic ideals is (pre)menses. The specific question about embodied experience that we will address is "What is the meaning of (pre)menses?" The way that a community values menses has been shown to affect the constitution of physical and psychological symptoms preceding and accompanying menstruation. I will call the formation of the latter as it shows itself in the postindustrialized West, the *psychosomatic constitution* of premenstrual syndrome (PMS). The difficulty physicians and women have agreeing upon a PMS definition indicates that PMS, rather than constituting an identifiable clinical entity, marks the simultaneous emergence of a transforming psychosomatic constitution and a set of psycho-socio-cultural factors, mutually transfiguring and transfigured by that emergence. That is, the way in which a community values women can affect a woman's bodily experience of the (pre)menses. I will show 1) that the way humans value menses affects how women experience (pre)menses; 2) that the way humans value (pre)menses is a function of historical and cultural conditions; and 3) the psychosomatic constitution of premenstrual syndrome appears with no determinate identity.

To begin, I would like first to distinguish two kinds of *valuation*.[6] They apply both to the (pre)menses discourse as well as the ascetic ideal discourse. There are valuations that bestow a preference for or prejudice against something. For instance, "Gilberto Santa Rosa is the greatest living salsa singer" makes a positive valuation and "Lying is wrong" makes a negative valuation. There is a second kind of valuation that interprets or evaluates the definition of an entity. For instance: "a kitchen utensil used to serve liquids" traditionally defines "ladel" and "a rational biped," human being. The first kind of valuation I will call *moral valuation*. It reflects a general cultural and/or particular individual bias for or against a certain constitution.

The second kind of valuation I will call *definitive valuation*. Definitive valuation defines a constitution. A constitution may have material characteristics, that is, a chair; or psychophysical attributes, that is, a disease; or theoretical qualities, that is, an economic theory. These can be provisionally categorized respectively as socio-physical constitution (material extreme), psychosomatic constitution (center),. or ideational constitution (immaterial extreme). The limits of these categories are not firm, but co-relating along a continuum of material and immaterial extremes. I will show that such constitutions are produced according to co-transforming moral valuations and definitive valuations. Such co-laboration points to an ontology described by the concept of dynamic non-dualism. Thus, moral valuation, definitive valuation, and psychosomatic constitution (the provisional category of the (pre)menses constitution) are relationally co-constituted.[7]

Turning to the topic of (pre)menses, our first consideration explores the relationship between the concept of (pre)menses and *moral valuation*—that is, how conventional valuations of (pre)menses relate to the way a woman experiences (pre)menses. "Menstruation," a term that does not overtly include the idea of disease, implies nonetheless, a series of negative values. Consider the following excerpt from Gloria Steinem's classic, "If Men Could Menstruate" (1983):

> [L]istening recently to a woman describe the unexpected arrival of her menstrual period (a red stain had spread on her dress as she argued heatedly on the public stage) still made me cringe with embarrassment. That is, until she explained that, when finally informed in whispers of the obvious event, she had said to the all-male audience, and you should be *proud* to have a menstruating woman on your stage. It's probably the first real thing that's happened to this group in years!
>
> Laughter. Relief. She had turned a negative into a positive. (Steinem 1983, 337)

Although this woman may have turned the negative into a positive, one thing is implicit: menses as something to hide, to feel shame about, is the norm in the millennial West.

Must such a negative moral valuation necessarily attach to menstruation? Steinem offers an interesting response. "So what would happen if suddenly, magically, men could menstruate and women could not?" (Steinem 1983, 338)

> Clearly, menstruation would become an enviable, boast-worthy, masculine event: Men would brag about how long and how much. Young boys would talk about it as the envied beginning of manhood. Gifts, religious ceremonies, family dinners, and stag parties would mark the day.
>
> To prevent monthly work loss among the powerful, Congress would fund a National Institute of Dysmenorrhea. Doctors would research little about heart attacks, from which men were hormonally protected, but everything about cramps. Sanitary supplies would be federally funded and free. Of course, some men would still pay for the prestige of such commercial brands as Paul Newman Pads, and Joe Namath Jock Shields—"For Those Light Bachelor Days. . . ."
>
> Street guys would invent slang ("He's a three-pad man") and "give fives" on the corner with some exchange like, "Man, you lookin' *good*!"
>
> "Yeah, man, I'm on the rag!" (Steinem 1983, 338)

Arguably, how we experience menses in the West is in part a function of how we value menses, that is, whether as "good" or "bad"—and the latter valuation, a function of how we value women. If men could menstruate, suggests Steinem, then menses would be affirmed, as is often the case with attributes linked with those at the top of a social hierarchy.

Of course, however, it is women that menstruate. Could this in part contribute to an overestimation of the supposed "negative" quality of PMS symptoms and the underestimation of possible positive aspects of their process? Consider the following premenstrual changes listed on a standard questionnaire used by some PMS researchers: angry outbursts, arguments, violent tendencies, anxiety, tension, nervousness, confusion, difficulty concentrating, crying easily, depression, food cravings, forgetfulness, irritability, increased appetite, mood swings, overly sensitive, wants to be alone (Rodin 1992, 54). We do not usually speak of positive experiences as "symptoms." Indeed, PMS symptoms, such as those listed, almost always harbor a negative value.

Do the symptoms above inherently involve a negative experience and therefore a derogatory evaluation? Might "food cravings" also be experiences of creative and passionate appreciation for food, and could "crying bouts" and "oversensitivity" be moments of insight and emotional perceptivity?

Only in recent decades have PMS studies begun to consider the possible positive aspects of female hormones. In one study, women were to report on positive and negative moods during the menstrual cycle. This experiment by Margie Ripper (1991, 25) asked women to evaluate over the course of a month, on a one-to-five scale, their moods (a five marking the best mood). Although women's moods were slightly lower during premenses than during other times, for the group as a whole, the measures of mood and performance remained above the midpoint, that is, above an "average" sort of mood throughout the menstrual cycle.

Perhaps most interesting, however, is that Ripper's experiment measures the positive emotions and performance-levels felt midcycle during ovulation. Premenstrual syndrome research has overwhelmingly sought to measure only negative symptoms (Ripper 1991, 25; Zita 1989, 190–201). Were it to measure the positive ones, writes Ripper, it might instead have been called Great Ovulation Elation Syndrome (GEOS) (1991, 26). For ovulation, which also occurs premenstrum, is for many women "a time of euphoria, creativity and heightened productivity" (Zita 1989, 191). It is significant that menstrual research has sought out apparently negative aspects of our menstrual cycle but has almost entirely overlooked possible positive ones.

The data above suggest that the physiological and psychological symptoms often linked with premenstrual syndrome may not involve inherently negative experiences. The perceived negativity associated with menstruation may, in part, be a function of how we as a society value women. I have also begun to show that even if there may be for some women some negative attributes of PMS, there may also be positive ones. For years, such positive attributes have been almost entirely overlooked from the medical and layperson's perspective. As Zita and Ripper indicate, the alleged midcycle peak of the premenstrum, "which is for many women a time of euphoria, creativity and heightened productivity" is commonly ignored (Zita 1989, 191). My point is not to take sides on this issue. Instead, what is important here is the process of the production of meaning implied by the ranging results of PMS studies. This is a process in which the moral valuation associated with the experience of the premenstrum appears to be co-constituted by the way a community values women—and by extension, certain reproductive processes symbolizing the feminine.

Recent research on Rabbinic and Christian reconstructions of gender and menstrual purity lends vivid support to the idea that a moral valuation of menses varies with context. In her book, *Menstrual Purity: Rabbinic and Christian Reconstructions of Biblical Gender* (2000), Charlotte Elisheva Fonrobert notes her surprise. When researching the concept of menstruation (*niddah*) in the foundational rabbinic literature, post-biblical texts of late antiquity (second-to-sixth century C.E.), she had not expected discussions about genital blood, colors and types of bloodstains, definitions of menstrual impurity, and of the sexual inaccessibility linked with impurity to be discussed with an absence of "embarrassment, shame or disgust" (Fonrobert 2000, 2). For Fonrobert, these are "feelings familiar to those of us who have grown up in the cultural context of the West, which allows mostly only euphemistic, hidden references to bodies and their messiness. The texts in Tractate Niddah might just as well be about zoology, astronomy, physics, or mathematics, judging by their tone" (2). Fonrobert analyzes research by Moshe Greenberg (1995, 69–79), who notes differences in the usage of *niddah* in biblical texts from in the post-biblical texts (17–18). References to *niddah* by priestly voices in biblical texts—most famous may be Leviticus 15—consistently link *niddah*, Moshe shows, with a sense of menstrual impurity (17–18). By contrast, post-biblical usage refers generally to "menstruant"—without the negative moral connotation (17–18). For instance, in the particular case of "*Mishnaic* Hebrew, the concrete sense 'menstruant' prevails" (Greenberg 1995, 76, qtd. in Fonrobert 2000, 18). Fonrobert's research lends support to Greenberg's. She finds that, "In the context of rabbinic texts, the meaning of the term *niddah* is primarily 'a woman who menstruates.' We do not find any texts or statements that indicate any valuation of the term, or that use the concept of *niddah* in any moralizing context, such as Exra or the prophet Exekiel might have suggested to the rabbis" (Fonrobert 2000, 18). Fonrobert finds, furthermore, that "the rabbinic literature does not anywhere indicate or allude to a practice of women's public segregation" (18).

Fonrobert's findings are surprising in the context of the two extremes Fonrobert says characterize contemporary debates about the role of women, past and present, in Jewish communities and debates about how the significance of menstrual impurity and ritualization of menstruation informs such roles (Fonrobert 2000, 15). One side of the polemic charges Jewish culture generally with sexism, and locates grounds for this view in the menstrual laws linking menses with impurity and prohibition of sexual relations between man and wife. The other side interprets positive results of the ritualization of menses, such as "its affirmation of women's physiology" (15). Fonrobert's findings are significant because they do not fall easily into either side. They indicate an absence of implied moral valuation in the use of *niddah* in rabbinic (post-biblical) texts (18). Thus, *Menstrual Purity* offers striking evidence showing the moral valuation of menses varying with context. With respect to the past and present roles of women in Jewish communities, we see this variance respectively in the two extremes of contemporary discourse, in biblical and post-biblical Rabbinic literature. This variance is significant because it indicates that moral valuation with respect to (pre)menses is produced according to a process of the production of meaning that is sensitive to the way a community values women and reproductive processes traditionally symbolizing the feminine. Fonrobert's conclusions about the use of *niddah* in post-biblical texts when compared to the two extremes characterizing contemporary debates or to biblical literature, show that the moral valuation associated with (pre) menses depends upon socio-historical context. This research lends support to the work of Steinem, Zita, and Ripper, which associates possible positive valuations with the premenstrum and implies that how a woman experiences her body during (pre)menses depends in part on how a woman/her society values (pre)menses.

If *moral valuation* of (pre)menses can be shown as a function of culturally conditioned factors, might a similar case might be made for a second kind of valuation—*definitive valuation* (definition)? Before we pursue the latter question, it will help to remind ourselves of the larger aim: *to show that moral valuation, definitive valuation and psychosomatic constitution are co-constituting.* Thus the desire to decisively define and identify a psychosomatic constitution—for instance, a so-called entity like PMS—is rooted in ontological misunderstanding. Just as moral valuation shows itself above to be conditioned, the process of *defining* premenstrual syndrome or aspects of it, will show conditioning by social and historical factors.

Medical research on premenstrual syndrome shows that physicians in the West have had a difficult time defining premenstrual syndrome. In a dominant so-called "Aristotelian" tradition in the West, *definition* (*horizmos*) has been understood as a formula articulating the essence of an entity.[8] In my view of Nietzsche's story, however, the veracity of such an essence may be rooted in thought structures that emerge in the West about the time of Socrates. I have been using the term "constitution" (i.e., ideational constitution) instead of "entity" to refer to a constellation of forces one defines. One simplifies the mobile and changing forces by naming them—in this case premenstrual syndrome—and attempting to define the name. Thus, that which begins as a metaphor for a complex and permutating constitution, becomes interpreted by many in the twentieth and twenty-first centuries as a signifier for a natural entity. By internalizing ideational valuations of PMS, men and women reinforce the appearance of a psychosomatic constitution of PMS.

The first physician to designate a group of symptoms related to menstruation as a clinical entity was Dr. Robert Frank in 1931 (Rodin 1992, 49–56). He used "premenstrual tension" to denote a series of symptoms arising the week prior to menstruation and ending the first day of menses. These included physiological experiences: headache, backache, abdominal and back pain, breast fullness and discomfort, weight gain, abdominal distension, fatigue and nausea; and psychological symptoms: depression, difficulty in concentration, nervousness, irritability, restlessness, and generalized emotional tension (51). The term "premenstrual syndrome" first emerged in 1953 when Dr. Katharina Dalton used it to describe the same symptoms delineated by Dr. Frank (51).

Notably, this early definition of premenstrual syndrome has been followed by a series of inconsistencies and ongoing disagreements regarding the character and definition of premenstrual syndrome, causing many to question whether such a "syndrome" exists at all. Take, for instance, the inability of medical researchers to agree upon the temporal location and duration of PMS in relation to menses. Some define PMS as the six days prior to menstruation and the first two days of menses. Others say it ends by definition at the onset of menses, whereas others agree that PMS need not be restricted to days preceding menstruation but may be "diagnosed *whenever there is a pathological variance* in the levels of estrogen and progesterone during the cycle" (Rodin 1992, 51). According to the latter, PMS includes approximately 17 days out of each cycle, more than half of every month, or, put differently, over 50 percent of a woman's reproductive life.

Moreover, there is not only inconsistency concerning the definition of PMS but also with regard to the symptoms attached to it. Not only do physicians disagree about which to include, but the symptoms themselves once selected as part of one's PMS definition often contradict one another. For instance one table includes the following opposed symptoms: insomnia, hypersomnia; anorexia, craving of certain foods; decreased concentration, paranoia (Zita 1989, 194).

Thus, in an attempt to isolate and firmly identify PMS, it seems that what one is isolating, that is, the supposed disease of PMS, is at least questionable if not entirely without determination.[9] By attaching the name "PMS" to certain symptoms, researchers imply that PMS is not evasive and amorphous however, but in fact an identifiable clinical entity. Earlier it was acknowledged that twentieth-century Western valuations of menses have been consistently perjorative. This is remarkable given the inconsistency surrounding the so-called "PMS entity" itself. If one is not certain what something is, how can one be certain if it's "good" or "bad?"

Some PMS researchers attribute this to an inability of biomedical researchers to test for PMS without importing into these projects culturally habituated assumptions about women and the menstrual process. Physicians Nada Stotland and Bryna Harwood (1994) agree that the definition of PMS and the negative valuation implicit in it affect the way women experience their bodies during the female cycle: "The perception of women as weakened by their reproductive functions, the reality of women's secondary place in society, and the concentration on pathology in medicine lead to research questions focused on negative experiences, reinforcing the circumstances that give rise to them. Research on positive concomitants of the menstrual cycle is rare" (191).

In the context of such research, PMS as a diagnostic category has many indications of being, at least in part, a socially constructed disease. This is not to say that it does not exist or that "it's all in women's heads." It is to say that PMS seems to emerge at least in part from an "ongoing and dynamic form of implied morality regarding women's preferred role in society as a wife and mother" (Rodin 1992, 55). Some studies show that women unfamiliar with the "western biomedical characterization of premenstrual changes do not experience what we call PMS" (55). Moreover, although some Western women experience premenstrual pain of various kinds, many do not. Evidence suggests that the *value* placed upon menstruation, a reflection of the value placed upon women, actually affects the way a woman experiences her body during menstruation. The equivocal clinical entity of premenstrual syndrome appears as a product of this reciprocating process. The production of meaning accompanying and shaping our experiences of and views about premenstrual syndrome coincide with the concepts of dynamic non-dualism and relation. Provisionally categorized as a psychosomatic constitution, PMS models overlapping planes traditionally viewed as separate: the plane of idea or psychology and that of matter or biology.

The nature and formation of PMS raises a political question, suggests Anne Figert (1996), and "any answers to these questions are political answers" (3). An interesting pair of court cases in Britain, in 1981, show why. In these two cases, women offered pleas of diminished responsibility due to premenstrual syndrome as a defense against murder allegations (Laws 1990, 189–190). Medical research from 1930 to 1985 attempted to identify PMS as a clinical entity and provided ammunition for a precarious new way of valuing women. An effect of attempts by such medical research to define PMS is that women have sometimes been evaluated as not fully capable of their normal activities or responsible for their actions. Especially in the late 1980s and early 1990s there was a popularized image of women—especially women in positions of authority—

with raging hormones. Anne Figert (1996) traces the possible emergence of this idea to "a public statement by Edgar Berman, Hubert Humphrey's physician during the 1968 Presidential campaign. Berman stated that he did not want a woman in a position of power because she would be subject to 'raging hormonal influences' each month . . . and 'Menstruation may very well affect the ability of these women to hold certain jobs . . . Take a woman surgeon . . . If she had premenstrual tension . . . I wouldn't want her operating on me'" (11).

Some believe that, compared to men's reproductive biology, women's reproductive biology is incommensurately stigmatized. Consider for example that "automobile insurance for adolescent males is considerably more costly than for people of any other age and gender. There seems to be an association between the surge of male hormones and the risk to people and property, but no 'testosterone-related conduct disorder of adolescence' has been proposed" (Stotland & Harwood 1994, 187). Consider also that women make up over 50 percent of the U.S. population but commit only about 10 percent of the violent crimes (Stotland & Harwood 1994, 192). In this light, the highlighting and stigmatizing of PMS seems all the more ironic, given that there is not a corresponding emphasis on hormonally related male behavior.

Significant here is not the matter of choosing a side in the moral evaluation of the PMS psychosomatic constitution nor of deciding whether to deny that PMS is a "reality" for some women. Of significance is the PMS example as a model and metaphor for a process of the production of meaning implied in Nietzsche's example of ascetic ideals.

PART III. CONCLUSION OF THE (PRE)MENSES EXAMPLE

The PMS example suggests that the way people value (pre)menses varies because the valuation is a function of socio-cultural conditions. (Pre)menses can attract and sustain positive valuations as well as negative ones. It also indicates that the definition of PMS and the psychosomatic constitution of PMS are themselves equivocal, transforming entities. In showing the indeterminacy of moral valuation, definitive valuation, and the constitution of PMS, my purpose has been to lend evidence to the following position: moral valuation, definitive valuation, and psychosomatic constitutions are co-constituted. They appear as related but not *relata*–not reified things. As co-constituting factors–moral and definitive valuations of PMS, and embodied experiences of PMS obfuscate a traditional distinction between subject and object and mind and body. Each factor is silmultaeously reciprocating and be reciprocated to. Each is thus not strictly distinct as a factor, as each is likewise not as a subject or an object.

It is such obscuring of subject and object that the grammar of a genitive phrase implies. Each term—"meaning" and "(pre)menses"—in the genitive phrase with which we started ("the meaning of (pre)menses") we could say operates both as subject and as object.[10] But if each "is both," then a distinction between subject and object is no longer clear. Here is made visible the opacity of subjectivity and objectivity per se. Ideational constitutions (moral and definitive valuations) and psychosomatic constitutions ([pre]menses) show themselves as co-constituting. Such formation accords with the concepts of dynamic non-dualism and relation.

Also significant is that by choosing the (pre)menses example, I begin to include a notion of biology—empirical statistics and ideas about life *as* sociophysical—in my concept of body. This is important, for many feminist philosophers of body have taken care to avoid biology—particularly reproductive biology—in their concept of body. This is a mistake, I think. It is a mistake not because biological discourses hold the key to a concept of body but because they offer an expansive set of data describing lived physiological and psychological experience. There is certainly precedence for this in the work of some phenomenologists, most notably that of Maurice Merleau-Ponty. Merleau-Ponty appeals to scientific data about human bodies and perception while also managing a phenomenological analysis that generally avoids a reductive biologism.

Reflection and studies on (pre)menses have been carried out by people not only in the sciences but also in the humanities. By choosing (pre)menses as the example to illustrate the process of production of meaning implied in Nietzshe's example of ascetic ideals, I am also able to elucidate Nietzsche's view of the similar limits and intersections shared among fields of knowledge. Such fields of knowledge—biological science, critical anthropology, and value theory—co-relate according to the concepts of dynamic non-dualism and relation. My interpretation of Nietzsche's story showing the relation between mind and body, and the formation of PMS with respect to its moral, definitive, and psychosomatic constitution, points to a mix that is neither a reductive monism nor a Cartesian dualism. The possibility of identifying disciplines discretely in any determinate way, like the possibility of identifying mind and body discretely is shown to be improbable. My analysis of Nietzsche's "Third Essay," like of his "Second Essay," discloses instead relations without *relata* (i.e., related things), and dynamism without dualism.

Using the example of (pre)menses, I have attempted to illustrate and lend support for Nietzsche's view of mind and body as a relation without related things. If the (pre)menses example serves to illustrate the dynamic non-dualism and relatedness of moral and definitive valuations of (pre)menses, and of variegated factors affecting the way a woman experiences her body during (pre)menses, then it seems that Nietzsche's story reciprocates by implying the concepts of dynamic non-dualism and relation. These concepts assist an analysis of the way meaning arises for and is ascribed to the contemporary phenomenon of PMS. The concepts of dynamic non-dualism and relation indicate that the reality of PMS is generally not a static function. They show that how a woman experiences her body during the premenstrum is in part a function of how society values women. They show, in other words, that "what" we discover PMS to be appears to be a function of the role, status and value we discover in, or assign to women.

The concepts of dynamic non-dualism and relation help to tell a story about PMS research from 1930 to 1985, about its implied morality about women, about its focus on certain perceived negative aspects of female hormones exemplified in this research, and about how it reinforces or even creates the very physiological and psychological symptoms it is bent on finding. Yet, one must also ask, is not this story—guided by concepts of dynamic non-dualism and relation—also "a smokescreen created to ensure the adherence of simple peoples to procedures that guarantee their physical survival" (Buckley & Gottlieb 1988, 23)? Indeed, and the forces constituting this tale of Nietzsche and embodiment, understood *as* smokescreen are crucial and noteworthy. They remain persisting factors in the production of meaning in the twenty-first century in the West.

The concepts of dynamic non-dualism and relation and the example of (pre)menses help to expose a non-dualistic model of embodiment wherein biology is co-participant in a field of overlapping planes. Women's biology, arguably not separate from women's mentality, seems to be a function of how we value ourselves, a function of how society values us, of how society values women. The (pre)menses example offers a non-dualistic model of embodiment that overtly includes a notion of biology. In addition, it shows why such inclusion precludes a reduction to biology.

The (pre)menses example also provides a metaphor for the process of the production of meaning implied in Nietzsche's example of ascetic ideals. If the (pre)menses example exposes reciprocity among moral valuation, definitive valuation, and psychosomatic constitution, the ascetic ideal example reveals reciprocity among moral valuation, definitive valuation, and ideational constitution. We can return to *On the Genealogy of Morals* and to the question of the meaning of ascetic ideals.

CHAPTER 4

Nietzsche's Ascetic Ideals *as* a Process of the Production of Meaning

PART I. THE MEANING OF ASCETIC IDEALS WITH RESPECT TO DEFINITIVE VALUATION

The title of Nietzsche's "Third Essay" (1887/1989), "Was bedeuten asketischen Ideale?" (KSA 5, 339), can be translated "What do ascetic ideals mean?" The German sentence assigns to the signifier *asketischen Ideale* a grammatical role as an agent giving forth or producing meaning. Indeed, the essay communicates that the concept of ascetic ideals is an agent for the production of meaning. The meaninglessness of suffering, not suffering itself, was the curse that lay over mankind so far—*and the ascetic ideal offered [bot] man meaning [Sinn]* [Nietzsche's emphasis] (GM 3, 28). Human beings suffer from meaninglessness, "*not* [from] suffering itself." The concept of the ascetic ideal presents itself to human beings as a fulfillment of a need for a purpose (*Sinn*)—a purpose for existence and for human suffering. The "ascetic ideal offered man meaning" [*bot ihr einem Sinn*] [Nietzsche's emphasis] (GM 3, 28).

At the beginning of his "Third Essay: What is the Meaning of Ascetic Ideals [*Was bedeuten asketischen Ideale*]?" Nietzsche refers to ascetic ideals in the plural. Many times in the essay, however, he refers to the ascetic ideal in the singular. In the first paragraph of the "Third Essay," Nietzsche lists multiple meanings of ascetic ideals (*asketischen Ideale*) and then unites these into one. He notes: "That overall the ascetic ideal [*asketische Ideal*] has meant so much, in this the basic fact of the human will expresses itself" (KSA 5, 339). Nietzsche refers to the ideal as a unity and as a plurality at different points throughout the essay. An inherent tension builds between a conventional tendency to view a constitution as an entity—as a unified particularity,[1] and an emerging sensibility to understand a constitution as a transforming constellation of forces.

Such conflict between the unity and the plurality of the concept of (the) ascetic ideal(s) can be illustrated by the several interpretations of the concept proposed in Nietzsche's "Third Essay" and by the logic of genitive phrases. Indeed, the different interpretations help to illustrate the obscuring of the agent-patient distinction significant in the structure of genitive phrasing. One recognizes that the ascetic ideal does not solely operate as an agent (or subject). The ascetic ideal also suggests in itself a role as a patient (or object). A theme in the "Third Essay" is that the ascetic ideal has many meanings.[2] Nietzsche shows these meanings emerging in relation to different types of persons—an artist, philosopher, or priest, for example—and in each case the ascetic ideal discloses a different moral and definitive valuation. "What is the meaning of ascetic ideals?—In the case of artists they mean nothing or too many things; in the case of philosophers and scholars something like a sense and instinct for the most favorable preconditions of higher spirituality; . . . in the case of priests the distinctive priestly faith, their best instrument of power, also the 'supreme' license for power" (GM 3, 1). The suggestion is that in any such context ascetic ideals not only exude meaning (agency) but also are imbued with meaning (patiency), that is, with moral and definitive valuation. Each human *interprets* the ascetic ideal. Thus, the ideal shows itself not only the producer of meaning but also the object of meaning, that is, of interpretation.[3] This dual role as subject and object conforms to a quality of conflating implicit in the logic of the genitive structure of Walter Kaufmann's translation of "Was bedeuten asketischen Ideale." He translates it: "What is the Meaning of Ascetic Ideals?" The genitive structure gives a twofold ambiguity. The "of" here invites both *ascetic ideals* and *meaning* to alternately be interpreted as the object or the subject of the genitive. This simultaneity as subject and object structure in genitive grammar, models the reciprocity named by the concept of dynamic non-dualism.

Conforming to the genitive structure is a related aspect of the ascetic ideal. This is the idea that the ascetic ideal has meant so many things to human beings (KSA 5, 339). In a manner paralleling the (pre)menses example, which meant differently to and was meant differently by disparate researchers, the ascetic ideal means and is meant differently from one context to another. In the "Third Essay" Nietzsche gives four examples of these various meanings. One occurs in the context of Wagner, another in relation to Schopenhauer, a third for philosophers generally, and a fourth with respect to the ascetic priest.

The (pre)menses example shows certain physicians, politicians, men and women of the twentieth- and twentyfirst-century industrialized West attempting to identify PMS. It shows that a generally agreed upon definition of this supposed clinical entity remains elusive. Similarly, Wagner, Schopenhauer, philosophers, and the ascetic priest definitively value the ascetic ideal differently. One reads the following about Wagner: "What does it mean, for example, when an artist like Richard Wagner pays homage to chastity in his old age? In a certain sense, to be sure, he had always done this: but only in the very end in an ascetic sense. What is the meaning of this change of "sense," this radical reversal of sense?" (GM 3, 2). The text shows Wagner's asceticism as a sensibility Wagner develops at the end of his life. Although he had always paid some attention to chastity, it is "only in the very end" that he does so "in an ascetic sense" (GM 3, 2). One infers that for Wagner, paying homage to the ascetic ideal in the form of chastity—that is, the will turning against itself, against sensuality—provides a means for Wagner to fight death. Thus, the ascetic ideal (as agent) offers for Wagner a means (Mittel) to a particular end: the preservation of meaning for life at the end of one's life. At the same time, Wagner interprets the ascetic ideal (as patient) as this means. Moreover, because the ascetic ideal helps to produce this particular end—a purpose for his life—it gradually comes to signify a so-called "meaning of life" for Wagner. One might say here that the ascetic ideal *means* this end. I mean "means" ambiguously here: the ascetic ideal serves as a vehicle—means to—the end *and* it signifies this end. "*[T]he ascetic ideal springs from the protective instinct of a degenerating life* which tries by all means to sustain itself and to fight for its existence. . . . The ascetic ideal is such an expedient [*Mittel*] . . . life wrestles in it and through it with death and *against* death; the ascetic ideal is an artifice [*Kunstgriff*] for the *preservation* of life" (GM 3, 13). The excerpt describes the ascetic ideal as a vehicle (*Mittel*) for fighting against death. It is a means. It also bears the second meaning, the act of signifying. For an artifice (*Kunstgriff*) is like a signifier. It is a text or design that has a particular value and significance according to socially conditioned rules. Its value and significance in this context is as a means for the preservation of life.

Wagner represents an interesting case for the genealogist. He manages to avoid allegiance to asceticism during most of his existence. This is an existence that the genealogist praises for its unusual ingenuity and courage evident in earlier works such as *Die Meistersinger*, which maintains the unstable balance between "'animal and angel'" (GM 3, 2).

Whereas Wagner appeals to the ascetic ideal in old age, Schopenhauer, according to the genealogist, indicates an association with it in the middle of his life and work as a philosopher: "Of few things does Schopenhauer speak with greater assurance than he does of the effect of aesthetic contemplation: he says of it that it counteracts *sexual* "interestedness," like lupulin and camphor; he never wearied of glorifying *this* liberation from the "will" as the great merit and utility of the aesthetic condition" (GM 3, 6). The moment of aesthetic contemplation for Schopenhauer, according to the genealogist, frees a human being from the binds of the will. According to the genealogist such freedom amounts to a turning against the will—a will turning against itself. As for Wagner, the ascetic ideal for Schopenhauer is a means to a distinct (if different) end—freeing oneself from the will. Also as for Wagner, for Schopenhauer the ascetic ideal comes to *mean*[4] its end. "'What does it *mean* when a philosopher pays homage to the ascetic ideal?'—here we get at any rate a first indication: he wants *to gain release from a torture*" (GM 3, 6). Schopenhauer, who applauds aesthetic contemplation for its ability to counteract "interestedness," the genealogist suggests, appears to have a hidden interest afterall. "[H]e wants to *gain release from a torture*" (GM 3, 6). The activity of contemplating art serves as a *means* to a certain end particular to Schopenhauer.[5] It *means* to Schopenhauer deliverance "from the vile urgency of the will" (Schopenhauer 1844/1969, 196, qtd. in GM 3, 6).[6] It *means* to Wagner forbearance against the steady draw of death. This *meaning* produced by the ascetic ideal (ideal as agent) reflects the meaning with which Schopenhauer and Wagner invest it (ideal as patient).

By "the meaning produced by the ascetic ideal" I mean the conventional, if fluid, meaning of the ideal as it presents itself to a perceiver such as Schopenhauer or Wagner. Such a meaning produced by the ascetic ideal and experienced by the perceiver we can call the ideational constitution of the concept of the ascetic ideal. The identity of such a constitution as ideational, however, is always already provisional. The forces of the ideal constituted by conventional institutions and exceeding the definitive and moral biases of perceiver, as experienced by the perceiver, are nonetheless relating to the perceiver's biases. Both Schopenhauer and Wagner have respective definitive valuation biases with respect to the ideational constitution of the ascetic ideal. Their respective definitive valuations of the ideal are constituted by forces reciprocally shaped by meanings produced among the dynamic community-based forces of the ascetic ideal. Once respective biases like those of Wagner or Schopenhauer are invested in the community-based ascetic ideal constitution (ideal as patient), such individual biases become constituent factors of the meaning produced by the community-based ascetic ideal (ideal as agent). The permutations of the ideational constitution of the ascetic ideal are results of forces that confound the logical distinction between the provisional constitution of the ascetic ideal "itself" and the experience of that ideal tailored by the biases of the experiencer. With respect to the definitive-valuation bias of the perceiver, the ideational constitution of the ascetic ideal shows itself operating simultaneously as agent and patient and vice versa. The definitive valuations of the ascetic ideal held respectively by Wagner and Schopenhauer likewise show themselves in relation to the provisional community-based constitution of the ascetic ideal, as both agents and patients.

In addition to singling out the particular philosopher Schopenhauer, the "Third Essay" speaks of philosophers in general, with respect to the ascetic ideal.

The essay considers three slogans the philosopher type associates with the ideal: poverty, humility, and chastity. These play a part in the "lives of all the great, fruitful, inventive spirits" (GM 3, 8). They form the "most appropriate and natural conditions of their *best* existence, their *fairest* fruitfulness" (GM 3, 8). For the philosopher the constitution of (the) ascetic ideal(s) provides an artifice for the possibility of being a philosopher, for creating philosophy. To do philosophy, according to the genealogist, the philosopher needs to distance oneself from chatter, busy-ness, newspapers, and politics (GM 3, 8). He needs to "be spared one thing above all: everything to do with 'today'" (GM 3, 8). Philosopher types revere "everything in the face of which the soul does not have to defend itself and wrap itself up" (GM 3, 8).

Ironically, the cloak of the ascetic ideal furnishes a wrap crucial to the philosopher's existence (GM 3, 10). "[I]n order to be able to *exist at all: the ascetic ideal* for a long time served the philosopher as a form in which to appear, as a precondition of existence" (GM 3, 10). The very possibility of philosophy, and the philosopher—"to be able to *exist at all*" depends upon the ascetic ideal, says the genealogist. In this light, for the philosopher, the ascetic ideal provides a means for existence per se. Again, as in the cases of Wagner and Schopenhauer, for the philosopher the ascetic ideal both *means* and is the *means to* his particular end. In the philosopher's case, this end is existence per se. It actively reflects the meaning invested in it by the philosopher, and so comes to signify this meaning. At the same time it acts as the expedient (*Mittel*) for attaining this end. It then conflates the distinctions of subject and object by simultaneously operating as subject and object and in two respects: 1) as patient reflecting the meaning invested in it by the philosopher and as agent offering a *means to* the particular end that it comes to *mean* or signify for the philosopher; and 2) as agent producing a community-based meaning of the ascetic ideal and as patient absorbing the experiencer's definitive valuation of the ascetic ideal that contributes to the elastic conventional meaning.

Precisely when the genealogist reveals the dependency of the philosopher on the "ascetic wraps and cloak," the genealogist introduces a fourth definitive valuation of the ascetic ideal, this time by the ascetic priest. The ascetic priest bears an interesting similarity to the philosopher. His possibility for existence also hinges upon the ascetic ideal. "The ascetic priest possessed in this ideal not only his faith but also his will, his power, his interest. His *right* to exist stands or falls with that ideal" (GM 3, 11). Like the philosopher whose existence as a philosopher relies upon the ascetic ideal, the ascetic priest's possibility for living as an ascetic priest—maintaining the "will," "power," and "interest" of the priest—depends upon the survival of the ascetic ideal (GM 3, 11).

Despite similarities, there is a key difference. Although both the philosopher and the priest depend upon the ideal for their existence, "the idea at issue here is the *valuation* the ascetic priest places on our life" (GM 3, 11). The priest, unlike the philosopher, categorically condemns "our life" and opposes it with the ascetic life—a life that turns against itself. "The ascetic treats life as a wrong road on which one must finally walk back to the point where it begins, or as a mistake that is put right by deeds—that we *ought* to put right for he *demands* that one go along with him; where he can he compels acceptance of *his* evaluation of existence" (GM 3, 11). Thus for the ascetic priest the ascetic ideal not only offers a means for the priest to exist, it also represents the apparently correct means, according to the priest, for not only himself, but all other human beings as well, to correct the mistaken direction of their lives. It is according to the ascetic priest not merely the only option for his or her own existence, but the only option for others too. In this respect, then, the ascetic priest "*demands* [*fordert*] that one go along with him; where he can he compels [*erzwingt*] acceptance of *his* evaluation of existence" (GM 3, 11).

Paralleling its role in the cases of Wagner, Schopenhauer, and the philosopher, the ascetic ideal *means* (it both expedites and signifies) for the priest his commanding evaluation of the ascetic ideal: *the will ought turn against itself!* It means that others interpret (definitively value) the ascetic ideal as the priest does and that the ascetic ideal *means* for others that which it means for the priest. It means that the meaning of the ascetic ideal, taken in its double-genitive sense, becomes or "is" unequivocal. In other words, according to the priest, it represents this determinacy to all who perceive and evaluate it, that is, as one should (*solle*) (KSA 5, 362). Consequently, all valuation of the ascetic ideal, whether moral or definitive, according to the priest, should reflect and gather itself in this determinacy. The ascetic ideal means: The will ought to turn against itself. The definitive valuation of the ascetic ideal says: The will ought to turn against itself. The moral valuation of the ascetic ideal concludes: The will ought to turn against itself. In the context of the ascetic priest then, the "ascetic ideal" is not ambiguously also "ascetic ideals" (plural). Instead, the concept of the ascetic ideal appears to congeal into stable, unified meaning.

For Wagner, Schopenhauer, philosophers, and the ascetic priest the ideational constitution of the ascetic ideal permutates and we have so far seen this with respect to definitive valuation. The ascetic ideal constitution as permutating is produced by a dynamism that conflates subject and object within and across the ascetic ideal constitution and the perceiver's definitive valuation of the ideal. The appearance of differing definitive valuations and of the permutating constitution of the ascetic ideal itself, suggest that the notion of the ascetic ideal as a coherent, identifiable conceptual entity, is fundamentally misguided. Such a proper concept appears not to accord with human experience; like the evasive clinical "entity" of PMS, the constitution of the ascetic ideal shifts according to the meanings invested in it by those to whom it appears. The "Third Essay" shows Wagner, Schopenhauer, philosophers, and the ascetic priest undergoing an ascetic ideal constitution that shows itself in two senses. In the first sense, as a community-based agent providing an experience for the perceiver, the ascetic ideal becomes the subject-genitive side of the double genitive phrase: meaning of the ascetic ideal. In the second, as patient of the perceiver's interpretive bias toward the community-based ideal, the ascetic ideal becomes the objective-genitive side of the phrase: meaning of the ascetic ideal. The ascetic ideal constitution shows itself as both the subject of an experience and the object of invested meaning (definitive valuation).

The ascetic ideal constitution for Wagner, Schopenhauer, the philosopher, and the ascetic priest not only shows itself as a function of reciprocally forming subjective and objective forces in the manner just indicated, but also in a second respect. The ascetic ideal is both a *means to* and comes to *mean* the definitive valuation with which the perceiver invests it. According to the "Third Essay," persons traditionally impose upon the ideal a hope for the ideal—that it will lead to a certain end. Conversely, one recognizes that the ascetic ideal invested with such a meaning begins to reflect this hope or meaning back to the interpreter. In this process of reflection, the ascetic ideal provides a means and bolsters one's will to that end. Note the active role here. The ascetic ideal now shows itself as an agent rather than a patient. It discloses its subjective-genitive side of the double genitive. In providing a means to the end, one can say that it means the end. And means is meant ambiguously here.[7]

If similar in some ways, the patterns suggested by the four cases of perception of the ascetic ideal constitution—considered above especially with respect to definitive valuation—also pose some disparity. The definitive valuation of the ideal by the ascetic priest, unlike those of the others, logically excludes the right to existence of competing interpretations. The priest *demands* that others accept his valuation of the ideal.

The priest can be said to represent momentarily in the text: presence, identity, or transcendence. One recognizes that the priest's existence "stands or falls with that ideal" (GM 3, 11). Additionally one sees that for the priest, the meaning of the ideal is an apparent unity. The ascetic priest, unlike the philosopher, Schopenhauer, or Wagner, does not wear the ascetic cloak like a mask. The ascetic life does not provide a means to a different end—for instance, persisting in the face of death or creating philosophy in a world hostile to philosophy. The ascetic life for the priest constitutes the core of his or her life, not its protective outer costume. For the priest the means and end of ascetic life are one: the ascetic life. Human life generally—as opposed to the ascetic life—says the priest, is "a wrong road on which one must finally walk back" (GM 3, 11). The priest excludes—theoretically at least—any other possible valuation of life, the ascetic ideal, and him- or herself. The ascetic ideal, for the priest, *means* the ascetic life and only that. The *meaning* (referent) of the constitution of the ascetic ideal appears, to the priest, unified—and appears as *means* to enact unification. It draws one toward the actualizing of an ascetic life. Because of this apparent unity, the ascetic priest understands oneself as self-identical with the ascetic ideal. In this sense, the priest fails to understand anything about the concept of relation. For the priest, beings are static *relata*—reified things, not relations of becoming. The priest understands oneself as a simple presence that is identical with the constitution of the ascetic ideal—reduced to and likewise misunderstood as a static identity.

For Wagner, Schopenhauer, and the philosopher, the ascetic ideal provides respective meanings: the will to life in the face of death; the release from a torture; the fairest fruitfulness for creating philosophy. Yet, these meanings do not mutually exclude one another. Neither encompasses nor self-identifies with the ideal. Each meaning shows itself as an enactment of a double-genitive reciprocity. Each appears constituted by a continual reshaping and gathering of forces, with developing, transformative—not unified, stable—meaning. Moreover, each permutation of Wagner's, Schopenhauer's, and the philosopher's view of the ascetic ideal recasts constitutions of forces and correlate significations within and across the various force fields constituting the ascetic ideal and the variegated definitive valuations of the ideal. In this sense, any perception of the ascetic ideal depends upon a dynamic constellation of forces that defy limitation to a coherent signification. To conceive the concept of the ascetic ideal as rational and proper, the "Third Essay" suggests, is to misconceive it.

The ascetic priest not only asserts himself as a unity in a text discrediting the plausibility of such unity, but also as an ahistorical figure in a text revealing in all probability, why figures are historically conditioned. The suggestion of an apparently "ahistorical" figure emerging amidst a realm of otherwise conditioned appearances, is a textual oddity that begs for explanation. "For consider how regularly and universally the ascetic priest appears in almost every age; he belongs to no one race; he prospers everywhere; he emerges from every class of society" (GM 3, 11). The ascetic priest appears "universally" in "almost every age." Does the genealogist intend to undermine a genealogical method supposedly rooted in the appearance of absence of absolute origins? Indeed, the genealogist implies that the ascetic priest arises in almost any epoch, race, or class, independently of social, political, and historical contingencies. How are we to understand a text supposedly rooted in a genealogical method that permits such an ahistorical figure? If genealogy is a method of inquiry that assumes that particular moral visions of the world, that is, of types like the ascetic priest, the aristocrat, the philosopher, or the slave are historically conditioned rather than ahistorically grounded and absolute—is Nietzsche's own genealogy exceeding its own initial limits?

To what extent is this assertion of ahistoricality constituting the ascetic priest able to maintain itself in the purview of the genealogist? I have shown that at least momentarily in the text, the ascetic priest understands him or herself, if mistakenly, as the ascetic ideal incarnate. The assertion of the priest's ahistoricality does not emerge from the view of the deluded priest, but from the perspective of the comparatively incisive genealogist. But immediately prior to linking the priest to ahistoricality (GM 3, 11) and also after it, the identity of the priest puts itself into question. It shifts to resemble that of the philosopher. One finds it worthwhile to ask, "What sort is the ascetic priest according to the genealogist? Is he (also?) the philosopher sort?"

Just before the genealogist attributes ahistoricality to the ascetic priest, the "Third Essay" attaches the following qualities to the philosopher-ascetic ideal dynamism. "[I]n order to be able to *exist at all: the ascetic ideal* for a long time served the philosopher as a form in which to appear, as a precondition of existence—he had to *represent* it so as to be able to be a philosopher; he had to *believe* in it in order to be able to represent it" (GM 3, 10). The philosopher had to *represent* the ascetic ideal and this meant he had to *believe* (*glauben*) in it. What is curious here is the idea that the philosopher has to *believe* in the ascetic ideal. Until this point in the ascetic ideal discourse, the genealogist associates the philosopher with the ascetic ideal as a sort of cover or costume that facilitates a philosophical existence—fruitful creativity and contemplation. The idea that the philosopher must believe in the ascetic ideal not only as a means but also as a sort of truth and end in itself, however, raises a new question. To what extent is one free to philosophize if one's beliefs are already committed to a certain first principle? The newly exposed meaning of the ascetic ideal for the philosopher differs from the earlier one. This new philosopher-ascetic ideal relation bears more similarity than the first, with the ascetic priest-ascetic ideal relation. In other words, this description of a philosopher begins to simulate the description of the ascetic priest.

In the following image, the philosopher and ascetic priest identities blend together: "To put it vividly: the *ascetic priest* provided until the most modern times the repulsive and gloomy caterpillar form in which alone the philosopher could live and creep about" (GM 3, 10). The ascetic priest provides the "caterpillar form," the costume or disguise "in which alone the philosopher could" exist (GM 3, 10). Moreover, the condition for assuming the disguise and for representing the ascetic ideal requires that the philosopher not acknowledge it *as* a disguise. To exist through representing the ideal, the philosopher must indeed believe the ascetic ideal (GM 3, 10). Here the philosopher's view of the ascetic ideal begins to superimpose itself onto that of the ascetic priest. The genealogist admits that "now . . . we behold the *ascetic priest*," and asks once again: "What is the meaning of the ascetic ideal?" As one recalls, at this stage the genealogist prepares for the emergence of another meaning of the ascetic ideal from the viewpoint of the ascetic priest: the identity of the ideal with the ascetic priest. Yet this supposed simple identity, belies a more complicated one. The genealogist asks: suppose such an incarnate will to contradiction and antinaturalness is induced to *philosophize* (GM 3, 12). The supposed simple identity of the ascetic priest begins to show more complex interests—characteristic of the philosopher. The genealogist's path shows neither the philosopher nor the ascetic priest maintaining respective selves or discrete views of the ascetic ideal. Just as the definition of PMS as identifiable entity continues to evade certain physicians and researchers, so too does a stable identity for the constitution of the ascetic ideal (and for constitutions generally), elude the genealogist.

The meaning of the concept of the ascetic ideal itself, the texts indicates, gathers and renders itself (ideal as agent), absorbs and is dispersed (ideal as patient) according to the logic of double-genitive grammar. Each of the various definitive valuations of the ideal developed in the respective contexts of Wagner, Schopenhauer, the philosopher, and the ascetic priest mean (ideal as object) and are meant by the ascetic ideal (ideal as agent). The genealogist taps into the conflation of agent and patient accorded the ascetic ideal. He/she points to it by exploring alternate ways the ideal *means* and is *meant* within and across constitutional planes. Thus, the view of the genealogist exposes at least three variegated reciprocating planes constituting the ascetic ideal. The first involves respective definitive valuations of the ideal held by of Wagner, Schopenhauer, philosophers, and the ascetic priest; the second involves the ascetic ideal constitution as *means to* (ideal as subject) and *meaning* (ideal as object) the definitive value given to it by the perceiver and the third involves the confluence of forces across the respective views of types of perceivers (i.e., the philosopher and the ascetic priest).

PART II. THE MEANING OF ASCETIC IDEALS WITH RESPECT TO MORAL VALUATION AND THE PERSPECTIVE OF THE GENEALOGIST

In Nietzsche's story, the constitution of the ascetic ideal is produced according to a function that includes the perceiver's definitive bias of the ideal. This function also includes the factor of moral bias. Wagner's, Schopenhauer's, the philosopher's, and the ascetic priest's respective experiences of the ascetic ideal imply their respective moral valuations of the ideal. Each generally gives the ascetic ideal a positive moral valuation. Each sees the ascetic ideal providing benefit—

if differing sorts—to one's life. If differences among definitive valuations of the ideal occur from one character type to the next, differences among moral valuations of the ideal occur between the four character types on the one hand and the genealogist on the other. Thus, as will be shown below, Nietzsche's "Third Essay" reveals not only competing definitive valuations but also competing moral valuations of the ascetic ideal. It sketches a process of the production of meaning that co-operates especially across the constitutions of moral valuation, definitive valuation, and ideational constitution. Thus, none of the borders of these constitutions is strict. If the moral valuation, definitive valuation, and the ascetic ideal constitution are dominantly of an ideational category of constitution, they also relate to—as shown in chapter 2—psychosomatic and socio-physical kinds of constitution. The planes of each constitution kind overlap. The ascetic ideal constitution, therefore, is not purely idea but has some confluence with bodily forces. The production of meaning with respect to the constitution of the ascetic ideal, one sees, coincides with the concept of dynamic non-dualism. By "dynamic non-dualism" I mean a field of overlapping planes, especially the planes of idea, psychosomatism and socio-physicalism.

One of several possible examples[8] from the "Third Essay" showing competing moral valuations of the ascetic ideal is the example of the genealogist's and Wagner's differing moral valuations. Wagner, says the genealogist, has "in a certain sense" always included an aspect of chastity in his work, but only in his old age does Wagner do so "in an ascetic sense." "What is the meaning of this change of 'sense,' this radical reversal of sense?—for that is what it was: Wagner leaped over into his opposite. What does it mean when an artist leaps over into his opposite" (GM 3, 2)? Of interest is the valuation the genealogist places upon the alternate qualities of the young and old Wagner. The genealogist attributes a positive value to the young more self-confident Wagner and a negative value to the old, more chaste Wagner.

In pursuing the question of the meaning of ascetic ideals, the genealogist implies that such meaning is in part a function of moral valuation. For instance, the genealogist considers the music Wagner would have created had he written an opera "Luther's Wedding." This would have "involved a praise of chastity [*Keuschheit*]. And also a praise of sensuality [*Sinnlichkeit*], to be sure" (GM 3, 2). This would have seemed "quite in order, quite 'Wagnerian'" (GM 3, 2). For there is no necessary antithesis between chastity and sensuality; every good marriage, every genuine love affair, transcends this antithesis (GM 3, 2) "[E]very good marriage," says the genealogist, "transcends" the opposition between chastity and sensuality. Luther himself displays the bold character of a master[9] who establishes "good" and "bad" values out of an inner strength expressing itself outwardly. For Luther "performed no greater service than to have had the courage of his *sensuality* . . . 'evangelical freedom,'" that is, to value as "good" that which most humans out of weakness and turning inward, value as "evil" (GM 3, 2). Like "every good marriage," Luther transcends the chastity-sensuality antithesis to realize "there is fortunately no need for it to be a tragic antithesis" (GM 3, 2). "At least this holds good for all those well-constituted, joyful mortals who are far from regarding their unstable equilibrium between "animal and angel" as necessarily an argument against existence—the subtlest and brightest among them have even found in it, like Goethe and Hafiz, one *more* stimulus to life" (GM 3, 2). Here we have before us the genealogist's view of the "well-constituted" human—somewhat unorthodox for nineteenth-century Europe because it describes the well-constituted in terms of a conflict—"unstable equilibrium between "animal and angel"—and this proving its secret: "its stimulus to life" (GM 3, 2).

In pursuing an answer to "What is the meaning of ascetic ideals," the genealogist looks at one case involving the ideal: "the case of Wagner." Interestingly, the genealogist interprets two very different Wagners emerging in this regard—the young self-confident Wagner and the ailing cowardly Wagner. The vibrant courageous Wagner, like "every good marriage" and "all well-constituted" (*wohlgerateneren*) mortals, embodies an "unstable equilibrium" of opposing forces: animal and angel, sensuality and chastity (KSA 5, 341). The genealogist imbues young Wagner who significantly is not yet beholden to the ascetic ideal, with the positive value of *wohlgerateneren*. That which is constituted by contradiction and an ability to transcend or thrive upon such internal conflict the genealogist positively values. Not only is there "no necessary antithesis between" such oppositions but "there is fortunately no need for it to be a tragic antithesis" (GM 3, 2). Instead such a tension is good, healthy—well-constituted.

Once the young Wagner becomes old, according to the genealogist, he leaps into his opposite: he becomes simple animal—swine, and only able to worship his opposite—the angel (or chastity) (KSA 5, 341). No longer hovering between two contradictories, the old Wagner pays homage to a delusory concept of the ascetic ideal, promising stability, consistency, and unity. In his new will-to-chastity, the older Wagner expresses an ascetic sense (*asketischen Sinne*). This is sensibility that implies allegiance to a permanence as a first principle and not—as Wagner's prior life experience would suggest, a first principle of impermanence. Such a predilection was absent in Wagner's younger character (KSA 5, 341). It constrasts sharply with the conflict-ridden, allegedly "well-constituted," young Wagner. According to the genealogist, Wagner's latter-day ascetic sense is simple, bad, ill-constituted.[10]

It is not difficult to deduce the genealogist's own moral preferences with respect to the ascetic ideal constitution. In the "Third Essay" the genealogist most often—but not always—views the ascetic ideal constitution negatively. In a genealogical study of moral preferences that implicitly denies possibility of a properly objective perspective, the genealogist cannot help but imbed into the text an additional layer of moral valuations—his or her own. This imposition requires factoring into the equation. What afterall is the meaning of "well-constituted," apart from the genealogist's valuation of the term with respect to a particular context? In what sense is that to which it attaches decisively *well*-constituted, that is, healthy or good independent of a contextual perspective that values it as such?

In my analysis of (pre)menses it becomes clear that the menstrual cycle receives a predominately negative moral valuation in the industrialized twentieth-century West. Some mistakenly interpret the premenstrum as a simple, precultural identity that presumably suggests another reification: premenstrual syndrome. The negative valuation attributed to (pre)menses shows itself dependent on the socio-historical conditions. As suggested earlier, some "symptoms" or experiences undergone during menses might actually be positively evaluated although habitually in the twentieth and twenty-first centuries in the West, they are not. Moreover, the constitution of (pre)menses shows itself transforming in relation to valuations associated with it. The views of menses so far conveyed in this discourse have revealed two opposed valuative poles: menstruation as debilitating or as productive; as negative or as positive. Yet one asks whether menses must be viewed in such valuatively clear-cut terms? Could one model the genealogist's own penchant for a more subtle balance of opposite forces? Could one see the menstruating woman as an "unstable equilibrium" (*labiles Gleichgewicht* (KSA 5, 341)) between the overly sensitive and the extraordinarily perspicuous, the pain and the ecstasy, the animal and the angel? Would she then mirror the genealogist's description of the "well-constituted" human—of a young Wagner, or a Schopenhauer? According to the genealogist, would this new balance between pain and ecstasy, good and bad in itself be "well"? Can some imagine this balance of opposites as does the genealogist—that is, as the mark of a 'well' constitution? Whether or not one does value the experience of (pre)menses in this way is not significant for my argument. What is important is showing the process of the production of meaning, suggested in Nietzsche's discourse of ascetic ideals.

We have examined the constitution of ascetic ideals with respect to definitive valuation. The definitive valuations of the ideal imposed by Wagner, Schopenhauer, the philosopher, and the ascetic priest differ from one another. They also contribute to the constitution of ascetic ideals, interplaying forces that reciprocate with the elastic community-based ascetic ideal. Implied in each of the four cases is a moral valuation of the ideal by the perceiver. Each morally values the constitution of the ascetic ideal or their modified brand of it, as a beneficial expedient to their lives. Although the genealogist typically counters their positive moral valuations with negative ones, (s)he sometimes counters her or his own negative valuation with a positive valuation of the ideal. The array of forces constituting the ascetic ideal with respect to competing definitive valuations, suggest a mobility consistent with the concepts of dynamic non-dualism and relations (without *relata*).

By exposing as part of the ascetic ideal equation, the genealogist's own conflicting moral valuation of the ideal, Nietzsche stirs up the already volatile set of ingredients. The dynamic structure of the stirred combination points toward dynamic non-dualism and relation. Significantly, this stirring and pointing in an essay about the production of meaning occurs in the process of producing meaning. The combination of ingredients shows the concepts of dynamic non-dualism and relation while enacting them. It shows dynamism not only within and across the two constitutions of definitive valuation and the ascetic ideal constitution, but within and across the three constitutions of definitive valuation, the ascetic ideal constitution, and moral valuation. The concept of the ascetic ideal, if in one sense the axis for showing this dynamism, in another sense is a metaphor for this process. Arising first, says the genealogist, to fill a gap in human experience, the ascetic ideal offers meaning for one's suffering. The ascetic ideal, suggests the genealogist, originates from a quest for meaning. In this sense, it can be said to symbolize the problem of meaning.

According to my analysis of Nietzsche's story, the problem of meaning examined by the genealogist shows the factors involved in perceiving meaning. These factors enact a process that is dynamic and relational. It is a process that magnifies the complexities of the concept of meaning. A factor in this complexity is moral valuation. The "Third Essay" indicates that the meaning of a positive valuation, (i.e., "well-constituted"), is conditioned by the viewer. In other words, items to which one might attribute a well-constitution are not objectively so. If one agrees that a necessary condition for the adjective "well" to possess a determinate meaning is that it attaches consistently to certain objects and not to others, then "well" has no such stable significance, according to the "Third Essay." Depending upon the perspective of the viewer, the so-called same item can be said to be good and bad (i.e., ascetic ideal, PMS). Also, the "same" constitution shows itself to be indeterminate—ever transforming. One can imagine that the adjective "well" or "well-constituted" will be attributed to an item at time "A" and again to the transfigured item at time "B," even though the item has changed. And this, the genealogist suggests, is a key factor in the deception. A dominantly favored metaphysics of permanence in the West indicates to subjects of the West, that experience gives rise to a number of discrete entities. Suggested in Nietzsche's discourse of the ascetic ideal, however, is that those apparent entities that many have attempted to define and categorize, appear not to be discrete, nor to accord with a metaphysics of permanence. Like experience, Nietzsche's story shows appearances as relational and changing. This view of the "Second" and "Third" essays of *On the Genealogy of Morals* differs from those of others who attempt to deduce what Brian Leiter calls "privilege readings" of Nietzsche's metaethical position (2000, 1). Two representatives of privilege readings are Richard Schacht (1983) and Philippa Foot (1991) (Leiter 2000, 1–2). They attribute to Nietzsche's moral critique an epistemic privilege over the morality it revalues. My view disconfirms privilege accounts by showing Nietzsche's genealogical project to emphasize not why any evaluative stance might be epistemically justified over another, but rather, a process of the production of stances, according to which epistemic privilege appears impossible.[11]

Since the rise of literacy in the West and the accompanying Christo-Platonic metaphysics—typically supposing a principle of permanence (i.e., forms, transcendent truth and reason founded on such stasis)—the application of a law of noncontradiction has become a deeply respected practice. The law presupposes that its object possess an aspect of permanence. Only if an object "X" has an at least momentary, if not persisting, identity *as* "X" does it make sense to assert that it cannot also at the same time and in the same respect be "not X." The assumption of such permanence, even while such permanence appears, for Nietzsche, *not* to be perceived phenomenologically (or experientially), has fueled in many a habitude of assuming that experience in fact yields appearances characterized by permanence. Put differently, a dominant metaphysics of permanence in the West remains visible in common practices such as the law of noncontradiction. Implied by the currency of the law, for Nietzsche, is the tendency among many to deny that appearances of experience show themselves as impermanent. The examination of the ascetic ideal in this chapter is significant because it brings to visibility the notion that phenomena of perception appear *as* impermanent. It displays a "radical empiricism that accepts the evident ubiquity of change in the universe without viewing it as issuing from, or tending toward, some being" (Cox 1998, 57).

Finally, although I have noted that the preferences of the genealogist are a factor in Nietzsche's genealogical study of morals, I have not discussed a parallel factor—the moral preferences of the *interpreter* of Nietzsche's genealogist. In this sense, just as the genealogist imposes a layer of moral preferences upon his genealogy of morals, so do I in my analysis of the genealogist's study. A genealogical narrator of morals[12] occupies a special relation to the discourse that he or she uncovers and promotes. In *On the Genealogy of Morals* the narrator is such a genealogist. He or she finds his or her own perspective and the discourse one unfurls subject to the limits and the biases of the valuative system inherited by their culture and its conventions. The genealogist lives amidst a moral and definitive system dominated by a supposed unified moral subject—with a will to knowledge, a will to simplicity and a will to unity. In *On the Genealogy of Morals* Nietzsche refers to this moral system as slave morality. Slave values attach "good" to behaviors, entities and activities that nourish and safeguard the "weak," "average," and "humble" person. By contrast the preceding moral system traced by the genealogist in the "First Essay: 'Good and Evil,' 'Good and Bad'" of *On the Genealogy of Morals* and out of which slave morality emerges, attributes "good" to actions and demeanors which nurture and support "strong," "single-minded," and "independent" persons. By dis-covering this moral history the genealogist discloses the conditioned nature of the moral system—a framework in which his or her own valuations and complex self-understanding are enmeshed. Subsequently, the genealogist discloses to oneself early on in *On the Genealogy of Morals* the impossibility of proving as absolute—as identifiable in a permanent way—any ideal. All ideals, all constitutions are only "good," are only "ideal" according to the conditioned perspective of the viewer viewing it as such.

The genealogist's recognition of the contingency of ideals[13] in the "First Essay" shows the genealogist's special relation to his or her own positing of preferences in the "Third Essay." The genealogist's predilection for the young Wagner who is "self-confident," who transcends the antithesis between "animal and angel," who maintains a delicate balance between these "contradictions," entails an idealization of these characteristics (KSA 5, 341). Yet because of the very terms of the unfolding discourse which has shown the inability of any valuational preference to persist absolutely, the genealogist's own positing of the delicate-balance-of-contradictories ideal—in this context an ideal undermining the dominant slave morality—on the text's own terms fails to wield an absolute sanction. It possesses instead a temporary prominence that the text will not be able to maintain. Thus the method used by the narrator—genealogy—reveals to the narrator (genealogist) that not only the history of the values that he or she unearths is conditioned, but also that his or her valuation of those values—at any given moment in the text—is conditioned too. This conditioning can be described having two levels: one general and the other particular. At the general level the genealogist's perspective is characterized by the discourse dominating the genealogist's heritage—that of the supposed unified subject. At the particular level his or her perspective reflects the specific path that the genealogist's own thinking and valuing forges. It derives and deviates from the more general discourse of the apparently unified subject carrying and covering over a preconceptual, pre-unified experience of self.

Similarly, my perspective in relation to the text, the genealogical narrator, and the (pre)menses model is that of a genealogist. My own valuations of these matters show themselves arising amidst a temporary setting of evaluative and ideational conditions. Yet just as the values, and constitutions transform, so will the appropriateness and relevance of my experience of the ascetic ideal be called into question and recoil. Indeed, my own valuations, likewise, are subject to the genealogical limits, valuational contingency, and cultural biases that I inherit at least as much as to those that I apparently choose. At the general level, the conventions I inherit are similar to those of Nietzsche's genealogist, that is, still steeped in a structure of thinking and valuing dominantly characterized by slave morality: will to unity, will to absolute truth and so forth.[14] At the particular level my valuations differ from those of the genealogist for they reflect the specific path and context of *my* conditioned thought. Also at the particular level, just as the genealogist's valuations may contradict each other during a process in which the conditions in *On the Genealogy of Morals* permutate, so too may my valuations. The genealogist's perspective, and my interpretation of that perspective, could likewise be spoken of in the language of storytelling. My story of the genealogist's story is made possible and limited by interpretive options and influences of the West in twenty-first century, just as Nietzsche's story, and moments within it, are made possible by the larger or immediate conditions, respectively.

Who then according to my story of the genealogist's, is the ascetic priest according to the genealogist? He or she is at once transcendental *and* indeterminate signifier, apparent simple self-identity and fractured historical figure. These contradictory renderings of the priest call attention to instability at the root of the process of production of meaning per se. The genealogist's own heritage—its language structures and imagery expressing a will to [a] unified concept of self and concepts—reveals at its basis co-influencing and conditioned perspectives. For the genealogist, the unlikelihood of the metaphysical unity willed by the ascetic priest points to conditioned arising and permutation of constitutions. The apparent incarnation of the ascetic priest as the ascetic ideal asserts itself as but one incarnation of the ascetic priest and one interpretation of the ascetic ideal. Both the ascetic priest and the ascetic ideal however, show themselves in the "Third Essay" exceeding this simple unity—relating to many more constitutions and valuations.

PART III. THE MEANING OF ASCETIC IDEALS: TECHNICAL CONCLUSION

Nietzsche's story about the production of meaning implied by the ascetic ideal suggests that traditional religious stories about ideals—grounded in a hope and belief in an absolute ideal, Origin or God—are not stories we have to tell. More important, suggests Nietzsche, these traditional stories seem more likely to obscure than to aid self-understanding. Nietzsche invites us to imagine how these traditional, official stories may do more to obstruct self-understanding than promote it.

Nietzsche's investigation beckons us to entertain his submerged story. In it both self-understanding and aspired-to ideals grow out of a lineage of localized, often unpredictable origins—origins borne on tides of motion, not on rocks legendarily thrown or ideally positioned by a biblical David or Peter. My story about Nietzsche's nontraditional story brings into focus what I call Nietzsche's concepts of dynamic non-dualism and relation. According to these concepts, human selves are both configurers of and constitutively of mobile constellations of forces. Of these constellations—perspectives, states of affairs, appearances, and situations are borne and die away.

The ascetic ideal and (pre)menses examples show that moral valuation, definitive valuation, and the ascetic ideal/PMS constitutions, respectively, are co-constituting. The factors of each example parallel each other as co-constituting. Moral and definitive valuations of the ascetic ideal and ideational experiences of the ascetic ideal parallel moral and definitive valuations of and embodied experiences of PMS. As co-constituting, these ingredients are relations without *relata*—without either momentary or persisting identifiable attributes to give an item proper identity. Such relatedness obscures conventional designations of subject and object, and mind and body. Ingredients are simultaneously issuing forth and being issued to. Ingredients, therefore, cannot be designated except provisionally; each defies determination in any strict sense, and by extension, as a subject or an object.

The grammar of genitive phrases models this defiance. Each term—"meaning" and "ascetic ideals"—in the genitive phrase "the meaning of ascetic ideals," like each term in the genitive phrase "the meaning of (pre)menses," acts both as subject and as object. If each is both, the distinction of "each," and of subject and object is not clear. In the case of the latter, ideational constitutions (moral and definitive valuations) and a psychosomatic constitution (PMS) show themselves as co-constituting. In the case of the former, ideational constitutions (moral, definitive valuations, and the ascetic ideal) show themselves as co-constituting.

Such a co-forming process coincides with the concepts of relation and dynamic non-dualism. Along these lines, the ascetic ideal constitution is categorized as dominantly ideational but is not reified as such. The constitution thus suggests the concept of relation. Each of the three planes of idea, psychosomatism and socio-physicalism show movement within and across one another. As participant in co-determining planes, the ascetic ideal constitution recalls the concept of dynamic non-dualsim. None of the borders of the planes is strict. The forces of each are co-operating. Such co-operation is more visible in the premenstrual syndrome example because not all of its primary factors are dominantly of the ideational plane. The PMS constitution appears to be dominantly of the psychosomatic plane, and the moral and definitive valuations of PMS of the ideational plane. The (pre)menses example shows biology as a co-participant in a field of planes not reducible to biology (or mere physiology) in any strict sense. Co-constituting that happens more dominantly across planes (the [pre]menses example) than within planes (the ascetic ideal example) helps make visible the non-dualistic model of embodiment that arguably informs co-constituting in general.

The concept of body that my story of Nietzsche's story develops includes a non-traditional notion of biology. By "biology" I mean empirical data and theory about life *as* socio-physical. Many feminist philosophers of recent decades avoid a notion of biology, especially reproductive biology, in their concept of body. Phenomenology in the tradition of Maurice Merleau-Ponty sets a precedent for including biological discourse in a non-dualistic model of embodiment. And although appeal to biological discourses provides no magic ointment for a concept of body, it does open one to expansive reserves of data recording the lived felt and imagined experiences of individuals. The practice of including a notion of biology in one's concept of body also avoids an unnecessary implication suggested by the avoidance of biological discourses. Conceptions of body that take care to avoid the data of biological discourse generally intimate that the lived experiences so documented pose a threat to a non-dualistic model of embodiment, unwittingly weakening their non-dualist concepts of body.

Nietzsche's investigation into the origins of the phenomena and experiences we configure and undergo invites us to imagine that the more traditional dualistic stories presumed necessary for self understanding and right comportment in the West appear not only not necessary but self-obscuring. It invites us to see the more traditional stories as constrictive renditions of our arguably broad and elastic beginnings—beginnings that include the emerging of a constitution of ascetic ideals in early preliterate cultures of the Mediterranean and Southeast Asia. I have shown variegated aspects of Nietzsche's discourse on the ascetic ideals constitution. These details show the ascetic ideals constitution for Nietzsche, to be a complex axis displaying an equally complex process of the production of meaning and showing so while in the process of such production. Moreover, these analyses of Nietzsche's discourse on ascetic ideals show the ascetic ideals constitution operating not only as an axis for such a showing but also in a second sense: as a metaphor for the problem of meaning per se.

The concept of ascetic ideals symbolizes a process of the production of meaning and shows the structure of this productivity. Ascetic ideals mean that meaning, in the face of its absence of a simple identifiable presence, creates and is created, transforms and is transformed. Meaning appears therefore simultaneously as subject and object—as an obscuring of these. Meaning gives rise to and transfigures meaning. One can say that *meaning means*.[15] The phrasing "meaning means," in hindsight, refers to the ambiguous subject and object identities discussed above. It also anticipates the grammar of certain preliterate cultures, especially of early Indo-European and Greek cultures discussed in chapter 8.

IV. THE MEANING OF ASCETIC IDEALS: GENERAL CONCLUSION

Nietzsche's tale suggests that the formative and contemporary influences of the concept of ascetic ideals are vast. As vast as he indicates them to be, an aspect of the concept of ascetic ideals remains familiar to us. According to this aspect, ascetic ideals often refer to a certain lineage or style of ideals, especially ideals promoting self-denial. Ascetic ideals as self-denial are recognizable to many of us. We experience frequently in our own quotidian lives and in those of others, ascetic ideals as self-denial. Such practices of self-denial range widely. Some are so familiar to us as to be barely visible or so glaring that we and people around us, cannot help but take notice.

Many practices of self-denial involve self-control. A paradigmatic example of self-controlled living can be found in the early practices of the historical Buddha. Before settling down with the less harsh routines the Buddha would eventually adopt, the Buddha is said to have disciplined himself so severely that he was near death. Day after day he would forgo sleep and eat only a few grains of rice and drink only drops of water. Eventually he denounced such discipline as too extreme and adopted a more moderate schedule of sleeping, eating, and meditation. Similar practices of self-control—Jews refraining from labor, business, and travel on the Sabbath,[16] Christians giving up a favorite pleasure-source for Lent, Muslims fasting sunrise-to-sunset for Ramadan, name some of the more prominent practices that can often become practices of self-denial and self-control. Many of us, if not adhering to them, are at least aware of the above century-old rituals. We are a culture highly tuned to self-denial as an ideal.

I remember the time my father's friend Doug rode his bicycle twenty miles to visit our family and to do some exercise. The ride covered lots of up-and-down chaparrel. We offered him lunch. "Oh no no no" he bellowed with tired breath coming from deep in his abdomen. "I haven't eaten solid food for over a month now." To cleanse himself he had entered into a modified fast. "All I eat is broth—broth from boiled fruits, from boiled vegetables . . . no meats, just broths—fruit and vegetable broths."

Doug's asceticism was not explicitly linked to any doctrinal belief in God. Asceticism or practices of self-denial, although often linked to religious traditions carrying a metaphysical belief in absolute truth and God, need not be. Although belief in God is the case for ascetic practices of the Judaic, Christian, and Muslim traditions, it is not for much of the Buddhist tradition. Unlike the others, much of the Buddhist tradition locates "truth" in impermanence. It tends to preclude a foundational origin resting in absolute truth and God. It follows then, that many of the Judaic, Christian, and Muslim traditions—by assuming God's existence—can attempt to deduce from that assumption a definite or proper concept of self. For much of the Buddhist tradition, by contrast, ascetic practices have been linked to a view of "no-self." More like Robert Lifton's description of a protean self (1993), reminiscent of Proteus—the Greek god of many forms—a Buddhist no-self is so-called because although an individual inhabits many forms some of the time, he or she inhabits none all of the time. For Lifton and for many Buddhists, a self's shape is fluid. It assumes one form and then another, and none permanently.

The concept of ascetic ideals appears like the shape of a Proteus, a Buddhist no-self and the experience of menses. It becomes many—if not just any—form(s); it seems never to remain one. The concept varies for each experiencer as does the experience of (pre)menses. In the "Third Essay" these respective meanings are conjured up by the likes of a cast of characters Nietzsche calls Wagner, Schopenhauer, the philosopher, and the ascetic priest. By hypothesizing a process of the concept's formation, Nietzsche's story suggests that the concept of ascetic ideals could never conform permanently to any imagined identity. Moreover, his story suggests that the process of formation of other concepts is similar to this one. No experience of human imagination, no concept, Nietzsche suggests—remains as or in a single form.

Nietzsche's choice of ascetic ideals as a concept through which to show the impermanence of identity of concepts has significance. The reason that humans fabricate ascetic ideals, Nietzsche's tale tells us, is to inscribe or find meaning in one's life, especially in one's life-suffering. The assumption of the existence of ultimate meaning for one's suffering typified in the West in a belief in God, suggests Nietzsche, motivates humans to practice self-denial, to attribute unity to the concept of ascetic ideals, and to project permanence onto other concepts as well. In the absence of a belief in God, in ultimate meaning for one's life, in one's suffering and in the permanent "proper" meaning of other imagined and felt experiences, life appears instead as unmeasured co-laboring.

Nietzsche says at the beginning of the "Third Essay" that "the ascetic ideal has meant so many things to man" (GM 3, 1) and he proceeds to seed a field with four growing and tangling variations of ascetic ideals. By reviewing the possible circumstances out of which the concept of ascetic ideals is born, Nietzsche not only indicates the impermanence of identity of the concept of ascetic ideals and of concepts generally but also that such impermanence coincides with a view of self as an immediate field created by and creating the perceiver and the perceived—a dynamic non-dualism, as I described in chapter 2. My view of Nietzsche's story shows how Nietzsche points to a way other than the official stories of Cartesian dualism and technological reason. If such conventional perspectives are made possible by assuming that the origin of one's felt and imagined experience includes the being of proper concepts, then by showing the absence of such being, my interpretation of Nietzsche discloses that the traditional Cartesian-Enlightenment perspective is in grave condition.

CHAPTER 5

Nietzsche on a Practice and Concept of Guilt

PART I. INTRODUCTION

Nietzsche investigates a cavernous overlapping of our embodied and reflected worlds not only with regard to concepts such as the ascetic ideal, but also with regard to personal preferences. In chapter 3 I argued that how we define a concept corresponding to a felt experience, such as menses, has historically varied in part due to how women have been morally valued. Although I focused on showing how the identity of the premenstrum has been difficult to pin down for among other reasons, the apparent intermingling of human imagined and felt worlds, one can also show that for Nietzsche, the identity of any one of a person's preferences is similarly difficult to determine. An example showing this is Nietzsche's moral preference against the behavior of guilt. Moral preferences like most appearances, situational conditions, beliefs, and feelings, Nietzsche's tale implies, arise, take ephemeral shape, and transpire. Fluidity, not rigidity, seems to characterize the appearance and contour of any of our preferences. Nietzsche's moral valuation of the practice of guilt does just this. It careens through several positions; and sustains virtually no characteristic permanence. This careening, as participant in a convulsing of forces and in helping to reveal the concepts of dynamic non-dualism and relation, points to the practice and concept of guilt likewise *as* a relation, not a reified identity.

PART II. TWO LEVELS OF MORAL VALUATION

To show this careening, it is worthwhile to first distinguish two levels of moral valuation operative in *On the Genealogy of Morals*. The two levels frame Nietzsche's moral valuation of the concept and practice of guilt. One is a general level. The genealogist recognizes it as the predominant moral system and sometimes calls it "slave morality." In the face of an all-powerful God, the pervasive morality positively evaluates the humble, meek and chaste. The other is a more particular level. It involves the specific for and against valuations made by an individual like the genealogist or myself. Valuations made on both levels, suggests the genealogist, are not absolute but conditioned.

Moral valuation at the pervasive level of "slave morality," Nietzsche's genealogy shows, indicates its historical contingency by revealing that out of which it grows. Its roots lie buried in a prior moral system sometimes called by the genealogist "master morality." This earlier morality positively evaluates that which its successor, "slave morality," condemns. Self-confidence, aggressiveness, and abundant strength are virtues. The genealogist's acknowledgement of the transvaluation from "master morality" to "slave morality" marks the first level—what I call the "general level"—of moral valuation. Furthermore, it depicts a moral tension at work throughout the "Second Essay."[1]

Moral valuation at the second level operates not on a sweeping but on a localized scale and marks the particular pro-and-con valuations made by an individual, for instance, the genealogist. An individual, although still immersed in a general cultural moral context—that is, "slave morality," will create his or her own path in relation to this context. In the case of the genealogist or myself who recognize the contingency of valuation, we may make a valuation that conflicts not only with the norm (slave morality) but also with one's own valuational bias. For instance, the genealogist who generally prefers to undermine the slave-moral norm, at times undermines his or her preference and boosts the norm. Significant in each case though, whether considering transforming values at the general level, or at the more local, personal level is that the process of the production of moral valuation, like that of definitive valuation, shows itself in Nietzsche's *On the Genealogy of Morals* not as reified, but relational. That which one sees transstructuring the definitive valuation of ascetic ideals and by extension, a concept of self—non-dual relations—one sees also with respect to moral valuation: the transvaluing of "good" and "bad." The revealed process of the production of valuation undermines the notion that human experience and supposed divine revelation are rooted in ultimate knowledge ensuring the consistency of a moral code or fixed definitions.

PART III. THE GENEALOGIST'S BIAS

We saw in chapter 4 that a moral value not only evaluates a constitution, be it sociophysical, psychosomatic, or ideational, but itself also constellates a constitution—

provisionally an ideational constitution. Each moral valuation enacts among other constitutions a reciprocity modeled by double-genitive grammar; each appears engaging simultaneously as so-called subject and object; and each obscures any supposed discreteness of subject, object, or itself as identity proper.

Chapter 5 shall illustrate that the genealogist's moral valuation of the disunified legal subject[2] emerges in two respects. First, it arises as a moral preference for the disunified subject illustrated by a calumny against those constitutions associated with its opponent, the supposed unified subject[3] exemplified by a practice of internalization and guilt. Second, it emerges as a discursive structure that intermittently interrupts the predominant moral favoring of the disunified subject. These sporadic interruptions create in the "Second Essay" a discursive structure that undermines the possibility of the unified subject. The genealogist's pervasively positive valuation of the disunified legal subject and the discursive structure periodically countering it disclose the concepts of dynamic non-dualism and relation.

The genealogist's predominate bias for the disunified subject can be seen in the genealogist's preference for practices of cruelty and the acceptance of animal instincts. Early in the "Second Essay," as the genealogist explores the relationship between cruelty and the making of a memory that would one day spawn a practice of guilt ("bad conscience"[4]), the genealogist morally favors this cruelty-inclusive culture over the guilt-ridden legacy that the genealogist inherits. "To see others suffer does one good," the genealogist writes, "to make others suffer even more" (GM 2, 6). The genealogist's avowal of physical punishment coincides with his or her understanding that humans are animals and therefore are characterized in part by instinctive irrationality. Naïvely, pathetically, sadly, indicates the genealogist, humans have come to condemn themselves for being, so to speak, what they are. "The darkening of the sky above mankind has deepened in step with the increase in man's feeling of shame *at man*. The icy disgust with life . . . I mean the morbid softening and moralization through which the animal "man" finally learns to be ashamed of all his instincts. On his way to becoming an "angel" (to employ no uglier word) . . . not only the joy and innocence of the animal but life itself has become repugnant to him" (GM 2, 7). The "increase in man's feeling of shame *at man*" and his growing disgust of "all his instincts," marks the rise of the reactive, apparently unified moral subject—a subject that rejects its instincts, "life itself," and yearns instead for the peace, purity, and otherworldliness of "becoming an 'angel'" (GM 2, 7). The genealogist's preference here is clear. The ascetic tendency of the so-called moral subject, the moral subject's idealizing "'the angel'" at the expense of this life, makes the genealogist sick.

We can recall from the "Third Essay" that the genealogist makes a similar valuation when he favors the young "well-constituted" Wagner. In contrast with him is the ascetic, aging Wagner, the "failed swine" (GM 3, 2). Wagner is no longer able to draw strength and exuberance from earthly life—the "unstable equilibrium between 'animal and angel'" maintained by the "well-constituted, joyful mortals" (GM 3, 2). He is a failure because he idealizes only one side of the opposition: that of the angel (GM 3, 2). The tension, the "'contradictions' that seduce one to existence" are denied as one wills only one side of the opposition—the beyond: its harmony, union, and accord (GM 3, 2). The old Wagner has turned against life.

The genealogist's disgust, in the "Third Essay," with the old Wagner, and in the "Second Essay," with the so-called unified subject, is provoked by a similar irritant: the appearance of life turning against life. The genealogist directs his or her allegiance instead to the disunified subject, to the young Wagner symbolizing such disunity, to the subject wrestling with contradictory animality.

The second illustration of the genealogist's moral preference for the disunified subject emerges in a discussion of the origin of justice—"the oldest and naïvest moral cannon" (GM 2, 8). The genealogist locates the roots of justice in the earliest systems of buying and selling. Its motto, "everything has its price" (GM 2, 8), as was shown in chapter 2, coincides with that of the legal subject. Unlike the "bad conscience" of the reactive subject, the conscience of the legal subject has not hardened under with the pressure of irredeemable penance. For the legal subject, a debt can be paid off. This formation of human experience and the context of forces co-laboring with it retain the opportunity to *turn* the meaning of a constitution towards a new end. The motto "everything can be paid for" reflects enacted practices. If a debtor has no access to the kind of constitution for which he or she is in debt—food, money, physical labor—the debtor and the creditor can *turn* something else (like cruelty) into the debt's equivalent. The legal subject's options for clearing the conscience coincide with his or her opportunity to *transform* (*verwandeln*) a spectacle of cruelty the debtor suffers into a pleasure the creditor enjoys and which not incidentally settles the debt (GM 2, 7).

The supposed unified subject, on the other hand, has willed away the opportunity to sustain or regain freedom, or more precisely, a clear conscience. By relinquishing the possibility of a liberated conscience, the unified subject also, says the genealogist, abdicates a perceptive aptitude that rests nearer to justice (GM 2, 11)

> The active, aggressive, arrogant man is still a hundred steps closer to justice than the reactive man; for he has absolutely no need to take a false and prejudiced view of the object before him in the way the reactive man does and is bound to do. For that reason the aggressive man, as the stronger, noble, or courageous, has in fact also had at all

> times a *freer* eye, a better conscience on his side: conversely, one can see who has the invention of the "bad conscience" on his conscience—the man of *ressentiment*! (GM 2, 11)

The legal subject who is "active, aggressive," and "arrogant" appears, to the genealogist more just than the reactive, unified subject. The relative nearness to justice of the legal subject is a matter of the conscience: "the aggressive man . . . also had at all times a *freer eye*, a better conscience on his side: conversely" it is "the man of *ressentiment*" who has "the bad conscience on his conscience" (GM 2, 11). The genealogist deems the legal subject's conscience "*freer*," "better," and also more just (GM 2, 11).

The above examples from the "Second Essay"—of instinct condemnation; and of the constricted unjust conscience of the unified subject—illustrate the genealogist's moral preference for their contrary: the disunified legal subject described by the genealogist as affirming life and retaining the option to live more freely and more near justice.

PART IV. THREE INTERRUPTIONS OF THE GENEALOGIST'S BIAS

If the genealogist's usual preference for the disunified subject displays consistency in the examples above, it does not in the essay's discursive structure. Although the genealogist predominately favors the early conscience and disunified legal subject throughout the "Second Essay," in the essay's discursive structure the genealogist interrupts his or her bias at least three times. Although these interruptions at face value undermine the genealogist's predominant bias and boost one's sentiment for the unified subject, at a more subtle level they support the genealogist's overall tendency to show the unlikelihood of such a unity. This is because these sporadic changes of preference reinforce the essay's discursive structure—a constellation whose changing relations undermine the idea of a consolidated subject. Thus, a parallel movement of transvaluation can be said to occur in terms of moral valuation so that even those values that locally applaud a unified subject nevertheless accommodate the text's larger form. This is a form of transforming relations, and so, undermines a coherent concept of self.

The first of the three interruptions occurs in section 16 in which the genealogist has decided to venture the first of two hypotheses about the origin of "bad conscience" (GM 2, 16). These I described in chapter 2 via two events: the so-called civilizing of human being by the blonde beast and the tribal community's turning its ancestor into a God. Section 16 depicts the first interruption (GM 2, 16). In it, the blonde beast instinctively confines human beings within certain civic walls and rules. Discouraged from venting its energy and will to power outwardly, the human animals release it inwardly—back against themselves. "[T]his animal that rubbed itself raw against the bars of its cage as one tried to "tame" it; this deprived creature, racked with homesickness for the wild, who had to turn himself into an adventure, a torture chamber . . . [T]hat," writes Nietzsche, "is the origin of "bad conscience" (GM 2,16). By associating the formation of "bad conscience" and the unified subject with an event that robs human being of liberty, the genealogist expresses again his or her overall bias against "bad conscience" and the supposed morally unified subject.

The genealogist nevertheless catches the reader by surprise when he or she shows that the transformation of conscience is not necessarily "bad." The genealogist does this by undermining his or her predominate bias against "bad conscience" which even in name bears the genealogist's bias against it.[5] "Let us add at once that, on the other hand, the existence on earth of an animal soul turned against itself . . . was something so new, profound, unheard of, enigmatic, contradictory. . . . From now on, man is *included* among the most unexpected and exciting lucky throws in the dice game of Heraclitus' "great child," be he called Zeus or chance; he gives rise to an interest, a tension, a hope, almost a certainty" (GM 2, 16). The genealogist notes some merits of the rise of "bad conscience." Its emergence brings forward something "enigmatic" "profound" "new" and "contradictory." One will recall that the "well" human being of the "Third Essay" is also contradictory. Is the genealogist suggesting here that the person of "bad conscience" is also "well?" It seems so. If only momentarily, the genealogist apprises us of the benefits of this arrival. Not only does the genealogist characterize it as "profound," "unexpected," "exciting," and "lucky," but also as worthy of comparison to Heraclitus (GM 2, 16), the only philosopher to earn praise from Nietzsche in *Twilight of the Idols* (TI, "'Reason' in Philosophy" 2).

Although Nietzsche could have created a genealogical narrator who remains consistent with his or her self-admitted preference for the early conscience and legal subject, a genealogist that occasionally changes views and undermines his or her predilection, manifests one guiding concept showing the way phenomena appear: relation. Phenomena, Nietzsche suggests, appear as transforming constellations of forces. The second moment during which the genealogist again surprises the reader by praising this new era of the subject, emerges in the discussion of the "ineluctable" disaster brought on by the wild blonde beast (GM 2, 17). In an instinctive burst of violence "blonde beasts of prey" entrap weaker forms of themselves—human animals—by creating civic obstructions. The genealogist links to the upsurge of "bad conscience" a coinciding emerging *beauty*. "[E]ventually this entire *active* 'bad conscience'—you will have guessed it—as the womb of all ideal and imaginative phenomena, also brought to light an abundance of strange new beauty and affirmation, and perhaps beauty itself.—After all, what would be 'beautiful' if the contradiction had not first become conscious of itself, if the ugly had not first said to itself: 'I am ugly'?" (GM 2, 18). "Bad conscience" provides the "womb of all ideal" which would bring "to light an abundance of strange new beauty . . . perhaps beauty itself" (GM 2, 18). "Beauty itself" implies not just any beautiful imagined phenomenon, but *the* idea of beauty. It suggests perhaps the most famous of such ideas, Plato's: the Idea of beauty as perfect knowledge in *Symposium* and *Republic*.

Plato's hypothesis of beauty appears sanctioned by its contrast to everyday earthly appearances. That Plato's cave allegory places the captives inside the earth accentuates the mortal situation. One finds oneself earthbound. The allegory can be read as a series of levels by which human beings ascend out of the earth and towards so-called correct knowing. A human being's will for a beyond—for a permanence and certitude foreign to earth—can be said to fuel this movement. The first step is begun by one's turning away from his or her everyday world of relations. When the prisoners are unshackled they turn around to see that the things that were previously revealed only through their shadows are now more "correct" than before. Arguably, the human of "bad conscience" is a descendant of this dominant reception of Plato, who allegedly likens earthly experience and the human instincts to shadows in a cave—infirm, inessential, and impermanent. According to this reading, Plato, unable to undergo human experience as it appears, says of the world, "It is impure," and of oneself, "'I am ugly.'"

Earthly life becomes the emblem of ugliness. And the antithesis—beauty—emerges, representative of all that mortal experience is bereft: certitude and permanence. Despite an implied link of beauty to the rise of a beyond-earth ideal, the genealogist applauds this development. "Bad conscience," with its ties to constancy and stasis, stimulates something novel: a "strange new beauty and affirmation, and perhaps beauty itself" (GM 2, 18). The "bad conscience" heralds an uncanny contradiction. It is "the ugly" but gives rise to "the beautiful." One recalls that the "well" human being in the "Third Essay" also displays contradiction. If the "bad conscience" does not maintain a delicate balance between "'animal and angel,'" as does the "well" human being, it nevertheless, suggests the genealogist, produces "contradictions that seduce one to existence" (GM 3, 2).

The emergence of the unified subject constellates a contradiction that births a new kind of beauty and the genealogist positively values it for its creativity. For "what would be 'beautiful' if the contradiction had not first become conscious of itself, if the ugly had not first said to itself: "I am ugly" (GM 2, 18)? As a soul "voluntarily at odds with itself" the reactive subject "makes itself suffer out of joy in making suffer" (GM 2, 18). It forms the ideal of beauty as an extension of its joy in recognizing itself as ugly. This contradictory being resembles the "failed swine" in "Third Essay" who worships only one side of its antithesis—

the angel beyond (GM 3, 2). Nonetheless, although in the "Third Essay" the genealogist devalues such a human, in this context of the "Second Essay" the genealogist elevates it. The novelty of its arrival elicits his or her excitement and enthusiasm. Thus, this valuation, as opposed to most in the "Second Essay," expresses an alternate moral valuation by the genealogist regarding "bad conscience." It shows the genealogist turning against his or her usual moral valuation that favors the disunified legal subject.

The third context in which the genealogist bestows a positive valuation on "bad conscience" and the reactive, morally integrated subject coincides with the genealogist's recollection of the emergence of the idea of God. The genealogist links this emergence to the rise of "bad conscience." In proportion to the success of the tribe and the subsequent feeling of indebtedness to one's ancestors, one would increasingly idealize one's ancestors—until finally they become God. The genealogist offers numerous pejorative moral valuations regarding this development. (S)he blames the advent of God for robbing human being of the possibility of clearing one's conscience. Because nothing can be turned into, used or meant as an equivalent of this debt, one finds oneself guilty to "a degree that can never be atoned for" (GM 2, 22). Under the debt's weight, one's conscience constricts and its liberty absconds (GM 2, 22).

Despite the genealogist's predominate bias against the practice of guilt or "bad conscience" and its relation to God, in this portion of Nietzsche's discourse there is nonetheless another momentary transvaluation of this preference. The "bad conscience," says the genealogist, is an "illness . . . but an illness as pregnancy is an illness" (GM 2, 19). Pregnancy creates. Thus, as in the preceding example of bad conscience's contradictoriness creating a strange new beauty, the genealogist expresses a positive valuation about the arrival of "bad conscience" precisely because it represents the creating of something new.

"Bad conscience" recieves at least three nods of approval from the genealogist in the "Second Essay." At the heart of these intermittent preferences is the prospect and hope that the genealogist attaches to the birth of something novel. Like the dice game of Heraclitus's "great child," the "bad conscience" gives rise "to an interest, a tension" and "a great promise" (GM 2, 16).

The genealogist's three interruptions of his or her general bias against the unified subject mark the second of two ways in which moral valuation shapes the discourse of the "Second Essay." The first occurs in a straightforward manner. This takes the form of the genealogist's repeated denigration of those constitutions associated with the unified subject, especially the practice of "bad conscience" or guilt. The second does so in a more discreet manner. The genealogist turns against his or her own general bias by suggesting positive valuations of the unified subject. Although on the surface this appears to undermine the genealogist's pervasive will to overturn the reactive so-called unified subject, beneath the surface it creates a discursive structure that gravely undermines the supposed being of the unified subject. By interrupting his or her moral preference, the genealogist points to Nietzsche's suggestion that moral valuation is—like definitive valuation—a changing relation, and can be said to avoid merely duplicating the reactive behavior it critiques. Rather, it enacts something like what Herman Siemens calls "agonal culture" (2001, 77). "Agonal contestation engages the antagonists in a complex interplay of mutal affirmation and mutual negation" and without the possibility of total annihilation or total victory of any one of the contestants (77). If the agonal strategy precludes absolute destruction of reactive subjectivity, it does not preclude the relative transvaluation of the contested dominant values.[6] The transformation of definitive as well as moral valuation constituting the texts of the "Second Essay" and "Third Essay," makes a viable unified subject all but impossible to assert. If as Nietzsche's *On the Genealogy of Morals* indicates, socio-physical, psychosomatic, and ideational constitutions participate in a reciprocal shaping process, then a unified concept of the subject appears incoherent.

PART V. CONCLUSION: THE PRACTICE AND CONCEPT OF GUILT AS RELATIONAL

In this chapter I have focused on the implications of the changing relations of moral and definitive valuations for a practice and concept of guilt. Nietzsche suggests that practices of guilt since the rise of Christo-Platonic values in the West, are pervasively assumed to be natural and corresponding to a properly identifiable concept of human nature. In the "Second Essay," Nietzsche discloses the flexibility of the genealogist's moral valuation of the practice of guilt. Nietzsche ties it to a discursive structure that renders a unified subject and an attending concept of guilt, untenable. In so doing Nietzsche's "Second Essay" entails certain consequences for the practice of guilt. Habits of guilt often believed from a perspective of the unified subject, to correspond to an identifiable concept of guilt are shown instead to be variable constitutions of conditioned forces. The condition highlighted above is that of preference. How one values a constitution (a concept of guilt, (pre)menses, or human subjectivity), my analysis of Nietzsche suggests, is constitutionally related to the formation and experience of the constitution for oneself and one's community. The genealogist's permutating preference regarding the practice of guilt, entails shifts to a formation and experience of a concept of guilt. This points to a concept of guilt *as* relational.

By investigating a lineage of the European practice and concept of guilt, Nietzsche continues to intimate that stories by renowned Anglo-European philosophers about the origins of human selves imply variant sorts of absolute origin. For such storytellers, acknowledging an absolute origin often becomes key to understanding one's supposed life purpose. For Descartes, one's origin is God, for Kant an a priori rational subject, for Hegel a free and ideal subject, for Marx a free material subject. In each of these oft-told stories—especially the "official versions" of them—the concepts of subject and world can be reduced to founding principles. This kind of reduction is precisely that which Nietzsche's story deflects. Nietzsche's story about the subject and the subject's experience of world, although revealing a reciprocity between subject and object, does not, like Hegel and Marx, reduce to an idealism or materialism. Nietzsche's narrative reflects and stimulates interpenetrating subjectivity and world. Subject and world exist there as co-generates of an immediate field of unmeasured interplay. Nietzsche reveals that human experience and its origins are identifiable provisionally not reifiably. And so, in exploring the concept of guilt in the West, Nietzsche discloses its variegating ancestry. He does this by implying what I call the concept of relation. Whereas in chapter 2 I show Nietzsche's concept of relation undermining a concept of a unified subject, here I illustrate it undermining a reified concept of guilt.

Amidst Nietzsche's sporadic interruptions of his apparent bias against the practice of guilt, Nietzsche implicitly underlines, I contend, a concept and practice of guilt as an array of relations. As a composite of relations, the concept of guilt, like the human subject itself, is also a relation and like virtually all relations, apparently indeterminate. Henri Bergson (1908/1991) describes conditions akin to the proposed idea of relation when he says,

> That there are in a sense, multiple objects, that one man is distinct from another man, tree from tree, stone from stone, is an indisputable fact; for each of these beings, each of these things, has characteristic properties and obeys a determined law of evolution. But the separation between a thing and its environment cannot be absolutely definite and clear-cut; there is a passage by insensible gradations from the one to the other: the close solidarity which binds all the objects of the material universe, the perpetuity of their reciprocal actions and reactions, is sufficient to prove that they have not the precise limits which we attribute to them. (209)

As seen in the case of the genealogist's valuation of guilt, Nietzsche shows that a person's moral valuation seems to be only provisionally identifiable, because of changing constellations of forces. How to analyze such protean fluidity remains a difficult task. By way of caution about too easily using the language of mobility and rest to describe parts of a whole, Bergson has something additional to say: "Though we are free to attribute rest or motion to any material point taken by itself, it is nonetheless true that the aspect of the material universe changes, that the internal configuration of every real system varies, and that here we have no longer the choice between mobility and rest. . . . We may not be able to say what parts of the whole are in motion; motion there is in the whole, nonetheless" (Bergson 1908/1991, 193). So too, my argument suggests, the concept of guilt appears only provisionally identifiable as a part—or whole—with parts in motion or rest. Nevertheless with the concept of guilt, as "in the whole" of lived experience (Bergson 1908/1991, 193), Nietzsche shows us, motion appears. It is a becoming that involves not only "self change" but "aspect change"—a change in the object and a change in the perspective from which it is considered (Cox 1998, 57). Christoph Cox notes that what "joins these two kinds of change is the *rejection of being* (57). Our concepts and our lived experience show themselves as what I call relations.

I began my introduction to Nietzsche's story by saying that Nietzsche's method of telling a story is to compile a number of substories or theories. Nietzsche rarely weaves a story without also submitting its pattern and premises to stress. I said that among Nietzsche's positions of play there is no official position—amidst his substories, no official story. And yet, in telling his story I emphasize two concepts: the concept of dynamic non-dualism and the concept of relation. I show in a number of scenarios why Nietzsche's story can be said to disclose both. It might appear then that my view of Nietzsche's story is contradictory—that in fact my reading of Nietzsche *does* attribute to Nietzsche an "official" if nontraditional story, organized by the concepts of dynamic non-dualism and relation. According to my view of Nietzsche's story, would Nietzsche say that meaning really produces itself in the non-dual, mobile, reciprocally shaping manner that the concepts of dynamic non-dualism and relations signify? Would Nietzsche lay to rest his habits of questioning and countering—at least with respect to these two concepts?

CHAPTER 6

Nietzsche, Metaphor, and Body

PART I. INTRODUCTION

As early as the first publication of *The Birth of Tragedy* in 1872, the concepts of metaphor and the body figure prominently in Nietzsche's thought. In "On Truth and Lying in a Non-Moral Sense," written in 1873, the roles of the concepts of metaphor and the body undergo a transition.[1] The change results from Nietzsche disbanding his belief in primordial unity and substituting for it a belief in apparent multiplicity. This development introduces a status for the concepts of metaphor and the body that will play itself out consistently through Nietzsche's subsequent writings and help answer the question of whether Nietzsche's story contradicts itself by both attempting to dissolve itself qua settled position, while relying on the structure provided by the concepts of dynamic non-dualism and relation. Before this development, however, and when Nietzsche still sustains a worldview presupposing a primordial essence, the goal for metaphor[2] is to best represent such a presumed world essence. Nietzsche's early writings adopt a hierarchy of metaphor with music at the top, spoken language below, and written language lower still.

Despite this hierarchal structure, *The Birth of Tragedy* anticipates Nietzsche's 1873 transition in its two senses of a loss of identity: (1) identity understood as "individuality"—lost when in Dionysian ecstasy the individual transposes oneself to become the other; and (2) identity understood as "essence of the world"—lost given that the individual has no access to it, but only to its best metaphor: music (Kofman 1993). In "On Truth and Lying in a Non-Moral Sense," Nietzsche does away with faith in primordial being and replaces it with a view of the world as an apparent multiplicity. This amounts to a loss of identity in three senses: as "individuality," as "world essence," and as "ground for a hierarchy of metaphors." With the dissolution of so-called right perception and truth, Nietzsche could be said to develop a new vision of "reality." In this vision, apparent reality is constituted by the bodily instincts as they interpret it (Blondel 1991, 206). The conscious intellect appears to be a mere appendage to the bodily instincts (206). This implied primacy of the body and the body's status as *interpretation*, is made possible as metaphor and through metaphor (205).

In chapter 6, I shall show that for Nietzsche, human experience as and through metaphor involves a two-fold principle of deception that points to a metaphysical limit. In accordance with this, Nietzsche does not say anything for certain (including for Nietzsche the link between apparent reality and three concepts developed in this book: dynamic non-dualism, relation, and metaphor), but he does not say nothing. In his writing this limit emerges as a dialectic of saying and unsaying.[3] For Nietzsche the interpretive status of the body as and through metaphor will also assist him in avoiding a reductive materialism[4] and idealism as well as metaphysical dualism.[5] We see this when Nietzsche links the metaphors of foundation and depth to the body (Blondel 1991, 212–213). In his underground investigations into the opaque body, Nietzsche shows the body to be the site that introduces *interpretation* into the world, moreover as the field that illustrates how hard it is to apprehend meaning (214, 245).

PART II. METAPHOR, THE BODY, AND *THE BIRTH OF TRAGEDY*

In the *The Birth of Tragedy*, the concepts of metaphor and the body operate differently in Nietzsche's thought than in his subsequent writings. They function within a schema of metaphysics that presupposes a primordial cause (being) (MW 31). In the context of *The Birth of Tragedy*, metaphor means the transposing of being into representation. It also means the transfiguration, self-dispossession, and ecstasy presupposed by this transposing of being.

If one understands metaphor in the sense of representation, one can see that *The Birth of Tragedy* and "On Music and Words"[6] establish a hierarchy of metaphors.[7] The body and the world exist for us only as images of the infathomable Dionysian primal cause (MW 31). We know that innermost nature "only in its metaphorical expressions" (MW 31). "Beyond that point there is nowhere a direct bridge which could lead us to it" (MW 31). Deviating from Schopenhauer's concept of Will (1844/1969), for Nietzsche the Will becomes the most general phenomenal form of an indecipherable Something we cannot fathom (MW 31). This most general manifestation we know through the life of instincts, "the play of feelings, sensations, emotions volitions" (MW 31).

In ways reflecting a Schopenhauerian metaphysics (Schopenhauer 1844/1969), Nietzsche distinguishes two categories of representation corresponding to two categories of symbolic representation. The fundamental category of representation is the sphere of bodily drives: pleasure and pain. They "accompany all other representations as a never-lacking fundamental basis" (MW 31). Secondary is the category of gesture that includes all other representation. The hierarchy between these two spheres reflects Schopenhauer's twofold character of the body as pleasure and pain, on the one hand, and as object of representation for a subject and conditioned by space, time, and causality on the other (1969). Nietzsche understands (and privileges) body as the first sense. Schopenhauer's body as object of representation is for Nietzsche a conception of the conscious intellect.[8] For Schopenhauer and Nietzsche the body as pleasure and pain is more primary because through it we understand the general form of all becoming and all willing. In this way the body and not the intellect shows itself fundamental in *The Birth of Tragedy* and "On Music and Words." Accordingly all representations that more directly appear in the form of feeling or emotion, are the more fundamental ones (MW 31). Just as this physiological form of representation is most basic in relation to other forms of representations,[9] so is the symbolic sphere corresponding to it more primary than the symbolic spheres corresponding to other sorts of representations (MW 31). The *tone* of speech symbolizes all degrees of pleasure and pain and the *gestures* of speech symbolize all other more secondary representations (MW 31).

Although the gestures of various languages are not universally understood, the sound and tone of iteration often are. Sound and tone, as the symbolic representation of pain and pleasure, symbolize a primordiality shared by all: the Dionysian primal unity. Out of the "tonal subsoil" develops "the more arbitrary gesture-symbolism which is not wholly adequate for its basis" (MW 32). Consonants and vowels without tone are simply "*positions* of the organs in speech, in short, gestures" (MW 32). A word coming out of a mouth is more than just a "position of the organs" by virtue of the tone-symbolism accompanying it. Tone as the basis of gesture-symbolism provides the echo of the pleasure-and-displeasure-sensations (MW 31). Just as the mere position of our bodily organs stand to their feelings of pleasure and displeasure" so the word built out of its consonants and vowels stands in relation to its tonal basis" (MW 32). In the sphere of language, speech conveys pain and pleasure through its pitch, rhythm, softness, or loudness and pacing. The written word, however, loses contact with this basic mode of bodily representation and is thereby less able to accurately disclose the Dionysian Will. "According to the doctrine of Schopenhauer, therefore, we understand music as the immediate language of the will. . . . On the other hand, image and concept, under the influence of a truly corresponding music, acquire a higher significance" (BT 16). Words, concepts, and images have a truncated significance but can be lifted to a richer level if accompanied by the appropriate tone.

Given this metaphorical hierarchy with respect to the symbolic representation, Nietzsche should have sung his creative works rather than written them.[10] Yet, having chosen the written word, it becomes clear that writing too has a hierarchy of symbolic spheres when it comes to expressing Will. As we will see, Nietzsche attempts to avoid the least appropriate of these: the conceptual language of philosophy.

It is possible to see why Nietzsche establishes a hierarchy of representational and symbolic realms. These correspond to a vision of the world in line with a slightly modified model of Schopenhauerian metaphysics. Regarding this hierarchy, tone is more primary than gesture. Both are subordinate to the most appropriate form for representing the innermost essence of the Will: music (MW 36). Thus, spoken language and written language are simply metaphorical transformations of the world's music. Moreover, because music best represents the Will, music will become a metaphor for the will (Kofman 1993, 8). Citing Schopenhauer Nietzsche writes, "We might, therefore, just as well call the world embodied music as embodied will" (BT 16).

This hierarchy of symbolic spheres also guides the appropriateness of subordinating the Apollinian to the Dionysian in *The Birth of Tragedy*. Lyricism objectifies Dionysian universality via Apollinian symbolism: images and concepts. The words of the poet can never fully capture the depth and truth of the world's melody (BT 17). For this reason, to use music as an attempt to "paint" the phenomena of Apollo represents the fundamental mishap of the theatrical arts after Attic Tragedy. Whereas the latter metaphorically transposes the Dionysian world melody via the Apollinian word and image, the *New Attic Dithyramb* paints the objective world in tones (BT 17). "Tone-painting is thus in every respect the opposite of true music with its mythopoeic power: through it the phenomenon, poor in itself, is made still poorer, while through Dionysian music the individual phenomenon is enriched and expanded into an image of the world" (BT 17). To ask music to depict the phenomenal world or speech to illustrate the written word is like asking a son to create his father (BT 33). It asks one to fashion the more primary after the less primary. Indeed, it is this clear delineation between the more basic and the less basic that establishes a hierarchy of metaphors in Nietzsche's 1872 work—*The Birth of Tragedy*—that distinguishes it from his subsequent thought. Although there will still be a hierarchy of metaphors in Nietzsche's later writings, there the distinctions will not be metaphysically grounded but based on the (non)foundational primacy of the body. The body is said to be a (non)foundational basis because the body transposes and is itself a transposition of multiple world forces. It is without identity proper. Although life and world will prove to be more than the body, Nietzsche's point will be that everything begins through the body.

Due to its renunciation of a hierarchy grounded in a transcendent or fundamental being,[11] Nietzsche's later work renders metaphysical identity generally untenable. Yet, even in his early work, *The Birth of Tragedy*, a loss of identity occurs in two senses.[12] In a basic sense, metaphor as it occurs in *The Birth of Tragedy* occurs with a virtual disintegration of individuality (Kofman 1993, 14). Nietzsche shows first and foremost that transposition happens to the individual. One has the ability to transpose oneself, that is, to almost break through the limits of individuality.[13] In such an act, the nearly dissolved individual becomes part of the unity of living things unified by Dionysus. The individual almost becomes the other; the many almost become the one. However, the possibility of metaphor rests in our not being able to fully transpose ourselves into a Dionysian unity. Part of the reason is that the "unity" of Dionysus has ever already been divided up. This is evidenced by our own individuation. "From the smile of this Dionysus sprang the Olympian gods, from his tears sprang man" (BT 10). Dionysus remains a god, dismembered. Metaphor presupposes this dismemberment and reconstitutes the god by symbolical transposition (Kofman 1993, 14). To look toward a rebirth of Dionysian unity is to conceive of "the end of individuation" (BT 10). The symbolical reconstruction of Dionysus coincides with the "almost shattered individual" that is kept from absolute dissolution by the Apollinian power of illusion (BT 21).

The loss of identity (the proper, the in itself, the patriarch) occurs in another sense, in the loss of an essential cause of the world. Although an individual presumes that one's experience or metaphors are derivative of such an "essence," one never has contact with the essence itself. In this manner, the metaphors lack a ground to enable them metaphysical identity. The "*one* primal cause" which lies beyond the supposed Will is "unfathomable to us" as a cause or essence (MW 31). This renders metaphors nothing more than transpositions of one another, that is, of the improper, inessential, and self-dispossessed. Nonetheless, they continue to refer to the "essential" by participating in a hierarchy that presupposes it. As part of a hierarchy presupposing an unfathomable in-itself, metaphors pretend to represent the inaccessible *one*, the other.

Music proves, for Nietzsche, the most adequate symbolical sphere for capturing the inaccessible essence of the world. As the best metaphor for the "in itself," music becomes the ersatz "identity" of the world (BT 16). All other representational and symbolic spheres become to greater or lesser degrees inadequate metaphors of music. Regarding symbolic metaphor, the written conceptual language of traditional philosophy lies the furthest from musical identity. Not only does it lack the expressive range of tone and sound, but it presents itself as non-metaphorical: as conceptual,[14] essential and proper. In denying the metaphorical character saturating it, it not only denies the world's primordial musicality and will of which it is derivative, but also its own absence of proper identity.

One sees here an inversion of the traditional "Aristotelian"[15] relationship between concept and metaphor. Whereas the so-called Aristotelian relationship subordinates the metaphor to the concept, Nietzsche subordinates the concept to the metaphor (Kofman 1993, 14–15). For the Aristotelian the metaphor is secondary in relation to the concept. One example is the categories of genus and species. The Aristotelian categorizes entities in the world in terms of genus and species according to a formula articulating the essential attributes of each genus or species.[16] Each category or concept distinguishes a logical or identifiable place. In *Poetics*, Aristotle (trans. 1941, 1457b6) describes metaphor as transferring to one thing "a name that belongs to something else." Thus, the name of one concept is "carried" from a proper place to a figurative one (Kofman 1993, 15). This may be from "genus to species, or from species to genus, or from species to species or on grounds of analogy" (Aristotle 1941, 1457 b7–9). In the case of Nietzsche, however, the Aristotelian notion of metaphor cannot be retained. Because of the inaccessibility of world identity, we are left instead with metaphor: the inessential instead of the essential, the improper instead of the proper, the son instead of the father. For Nietzsche, human experience yields only imagined, apparent identities constituted by metaphors. And so Nietzsche judges the concepts metaphorical.

With the visible centrality of metaphor in *The Birth of Tragedy*, one might be tempted to believe that the concept of metaphor itself usurps Dionysus's role as unifying principle. But the concept of metaphor instead is only a metaphor for the human epistemological condition. Nietzsche's reference to the "son wanting to create his father" establishes a metaphor for metaphor (MW 33). The son cannot create the father, nor can metaphor create the fundament. This metaphor for metaphor establishes a devaluation of metaphor per se (Kofman 1993, 15).

The Birth of Tragedy and "On Music and Words" recognize music per se—and not metaphor—as the best representation of the in itself. They also establish a hierarchy of representational and symbolic systems. Music's privilege presupposes that which it artificially reconstitutes—a Dionysus prior to dismemberment. We are left with on the one hand, a fragmented Dionysus and, on the other, an unfathomable primal unity.[17]

PART III. METAPHOR AND THE BODY AFTER *THE BIRTH OF TRAGEDY*

Metaphor and the Body in "On Truth and Lying in a Non-Moral Sense"

The short essay "On Truth and Lying in a Non-Moral Sense," written in 1873,[18] marks a transition in Nietzsche's thinking. Nietzsche will shed his faith in a

Schopenhauerian-like Will and in music as the best metaphor for such being. This change, as we saw, is anticipated in *The Birth of Tragedy*. We saw this in the two senses of the loss of identity. With no access to a possible world essence, and with metaphor itself presupposing a fragmented Dionysus in need of reconstitution through the arts, Nietzsche has little basis for naming music as the best metaphor. If one cannot access primal being, one cannot know whether it exists, let alone its character, or that which would be a best metaphor for it. In "On Truth and Lying in a Non-Moral Sense," metaphor no longer presupposes a Schopenhauerian primordial Will nor does it assume an opposition between ersatz-identity (music) and nonidentity (all other representation). Once Nietzsche discards the residual Schopenhauerian metaphysics, metaphor instead presupposes a world of multiple forces that remains unknowable, and operates amidst a body whose multiple forces remain opaque. So-called human reality is no longer a series of better or worse copies of world Will, but an anthropomorphic interpretation designated as metaphor and through metaphor.

"On Truth and Lying in a Non-Moral Sense" makes a significant break from *The Birth of Tragedy* by replacing a metaphysics privileging music with an anti-metaphysics and an "*artistically creative* subject" (TL 148). There is no "correct" perception, no cause, no truth, no in itself of things. There are instead interpretive carry-overs, metamorphoses, and transfigurations (TL 146, 148). Nietzsche's subject as artist and creator, if unique in some of its details, certainly is not unique in its general theme. Instead, typical of nineteenth-century romanticist principles, it emphasizes an implied artistic nature that pervades human experience. Such romanticism itself shows influence from Kant's transcendental categories of the subject and Kantian subjectivity when taken to a variant extreme. If the human subject shares an intersubjective objective world with other humans (but not with other creatures)—made possible by common ways that human organs organize external forces into intelligible forms—with a small conceptual leap, the notion of a shared objective world for humans can be transformed into a brand of perspectivalism: truth resides in the unique experience of the individual (i.e., Wordsworth, Coleridge). Nietzsche's neo-Kantian[19] subjectivity emphasizes the singular experience of aesthetically creating individuals as opposed to the shared objective categories of an artistically distinct human species. The attribution of creativity shifts from the individual as human to the individual as individual.

Within Nietzsche's own corpus, as Gregory Moore (2002, 85–111) has shown, a concept of *Kunsttrieb* (artistic drive) remains—if undergoing at least two metamorphoses—throughout his work. The first variant coincident with Nietzsche's *The Birth of Tragedy* describes Schopenhaurian artistry in the form of an underlying will driving production of representations. The second (to be described in this section in terms of an impulse to metaphors) has shed the Schopenhaurian metaphysics. The third develops the second and begins to substitute for the language of "artistic drive" (*Kunsttrieb)* and "metaphor drive" (*Triebe zur Metapherbildung* [TL 150]), the language of "will to power" (*Wille zur Macht*).

Nietzsche's concept of *Kunsttrieb* and its variegated developments follow certain previous incarnations of the term. Gregory Moore (2002, 89) locates the first use of the term *Kunsttrieb* in Hermann Samuel Reimarus's *Allgemeine Betrachtungen über die Triebe der Theire, hauptsächlich über ihre Kunst-Triebe* (1760), a book that "originally explained certain spontaneously creative behaviour observable in animals, referring to those instincts, for example, which prompt the bird to build its nest or the beaver its dam—this is the sense in which Schopenhauer, for example, employs the term in the chapter entitled 'Vom Instinkt und Kunsttrieb' in the second volume of *Die Welt als Wille und Vorstellung*" (Moore, 2002, 89). A brief history of the term, traced by Moore, shows *Kunsttrieb* soon also to be used in the context not of an impulse to above described animal technologies, but to human production of fine art (89). Reginald Snell's translation of Friedrich Schiller's *On the Aesthetic Education of Man*, for instance, translates *Kunsttrieb* as "aesthetic artistic impulse" (Schiller 1867/1994, 127 qtd. in Moore 2002, 89). Moore's research shows that both meanings of the term become combined in the work of the nineteenth-century German biologist Ernst Haeckel. "Haeckel's use of the term unites both meanings by giving the idea of the *Kunsttrieb* an evolutionary twist: human artistry is simply a more refined expression of the same primordial creative instincts which all organisms possess to a greater or lesser degree" (Moore, 2002, 89).[20] Thus, it is with the available currency of earlier uses of the term *Kunsttrieb*, that Nietzsche's own adoption of it in *The Birth of Tragedy* and the subsequent development of it in terms of *metaphertrieb* in "On Truth and Lying in a Non-Moral Sense" emerge.

For Nietzsche, the experience of the artistically creative subject described in "On Truth and Lying in a Non-Moral Sense" is constituted by an ongoing series of transpositions (*Übertragungen).*[21] The carrying-over from world to body and body to intellect involves a translation of forces from one sphere into another. "[F]or between two absolutely different spheres, such as subject and object are, there is no causality, no correctness, no expression, but at most an *aesthetic* way of relating, by which I mean an allusive transference, a stammering translation into a quite different language" (TL 148). No longer committed to any so-called right perception or best metaphor presupposing an underlying will, Nietzsche says the various layers of human experience are determined by an artistically creative subject according to an "*aesthetic* way of relating." The movement from external world to body or body to intellect is a creative one in which one transcribes the forces into values appropriate to the body, sense perception, and the intellect. The emphasis in "On Truth and Lying in a Non-Moral Sense" is more a devaluing of the intellect than a privileging of the body. Nietzsche repeatedly characterizes the intellect as misleading regarding itself and as deceptive in general. "The arrogance inherent in cognition and feeling casts a blinding fog over the eyes and senses of human beings, and because it contains within itself the most flattering evaluation of cognition it deceives them about the value of existence" (TL 142). The intellect (cognition) "shows its greatest strengths in dissimulation" (TL 142). "This art of dissimulation reaches it peak in humankind, where deception, flattery, lying and cheating, [and] speaking behind the backs or others . . . is so much the rule" (TL 142).

That experience for the artistically creative subject is constituted by interpretive metaphor and not "proper" concepts marks our self-deceiving nature. Perception is the creation of a picture or image (*das Bild*). "The stimulation of a nerve is first translated [*übertragen*] into an image: first metaphor! The image is then imitated by a sound: second metaphor!" (TL 144). By calling perception a type of metaphor, Nietzsche shows that the transpositions from external world forces to one's percepts, and from one's percepts to words, sounds or concepts, are not replications of an object and its so-called essence. Indeed the object of transposition has no proper essence: objects are from their start metaphors—that is, without originary identity. All are "allusive transferences" (TL 148), that is, translations and carrying-overs of one another—of the improper and self-dispossessed. If metaphors are deceptive it is because they are not equal to their referents; they transform them.

> Every concept comes into being by making equivalent that which is non-equivalent. Just as it is certain that no leaf is ever exactly the same as any other leaf, it is equally certain that the concept "leaf" is formed by dropping these individual differences arbitrarily, by forgetting those features which differentiate one thing from another, so that the concept then gives rise to the notion that something other than leaves exists in nature. (TL 145)

Likewise, each percept or picture of an individual leaf is a metaphor, a transposition of world forces into a human percept. This system of transposition describing the relation between external world forces, bodily perception and intellectual concept indicates for Nietzsche factors circumscribing human experience. Our drive to create metaphors marks "that fundamental human drive which cannot be left out of consideration for even a second without also leaving out human beings themselves" (TL 150–151). The body and intellect produce metaphors—carry-overs, allusive transferences, translations. A metaphor does not replicate any supposed "proper" aspect of its object but stands in for the object as it interprets it.

Such interpretation demands what one might call "creative artistry" on the part of the human organism, and such artistry, Nietzsche suggests, is not only an overlooked characteristic of our lived experience but a forgotten aspect of our ancestry. Humans are artistically creative subjects that forget their aesthetic nature (TL 148). In conjunction with such a *Kunst-* or *Metaphertriebe* Nietzsche acknowledges a Kantian-like subjectivity. His point, though, is not to argue for the so-called objective knowledge of subjectivity but to devalue such supposed knowledge by generalizing the subjective principle[22] to all living and nonliving things. Objectivity would vary "if we ourselves could only perceive things as, variously, a bird, a worm, or a plant does, or if one of us were to see a stimulus as red, a second person were to see the same stimulus as blue, while a third were even to hear it as a sound, nobody would ever speak of nature as something conforming to laws" (TL 149).

In a certain sense, therefore, Nietzsche denies that as subjects we have anything one might call "knowledge." Such knowledge would presuppose identity, essence, and truth regardless of our being a human—as opposed to another subject kind. That one often speaks as if one does have knowledge is because "man forgets himself as a subject" (TL 148). A human being transforms everything into its own subjective forms of space and time and from the "rigour and universal validity" of the latter, deduces the laws of nature (TL 149–150). The laws of number are borne by these very forms and "number is precisely that which is most astonishing about things" (TL 150). "[I]f we are forced to comprehend all things under these forms alone, then it is no longer wonderful that what we comprehend in all these things is actually nothing other than these very forms" (TL 150). Rather than yielding an objective world that is knowable, the artistically creating subject generates a world according to supposed knowledge, which deceives. By presupposing forms that we produce ourselves, truth is tautological. "If someone hides something behind a bush, looks for it in the same place and then finds it there, his seeking and finding is nothing much to boast about" (TL 147).

The possibility of such tautological aesthetic creativity presupposes a principle that replaces the Schopenhauerian metaphysics of *The Birth of Tragedy*. This is the principle of a freely creating force. Nietzsche's generalization of Kantian subjectivity resulting in his perspectivalism is made possible by his assumption of such a creative freedom, described as a "meditating force" between "two absolutely different spheres, such as subject and object are" (TL 148). This functions as a basic force making translation (*Übertragung*; *Übersetzung*) possible between the world of absolute chaos and the subject. Between these "there is no causality, no correctness, no expression, but at most an *aesthetic* way of relating, by which I mean an allusive transference [*Übertragung*], a stammering translation [*Übersetzung*] into a quite different language. For which purpose a middle sphere and mediating force is certainly required which can freely invent and freely create poetry" (TL 148).

Interestingly, Nietzsche's assertion of a freely creating force underlying his attribution of subjectivity to all things (perspectivalism), says something about the world beyond the human subject. Precisely because of such human subjectivity Nietzsche says we do not know anything for certain, particularly about the world of multiplicity beyond our subjective limits. And yet, both in spite of and due to such perspectivalism, Nietzsche posits the metaphysical principle of a freely creating force. This principle supports his attributing subjectivity to creatures generally, but at the same time, such subjectivity prevents him from attributing, in any dogmatic sense, the existence of such a force to the external world beyond. Borrowing Blondel's language, we can say that Nietzsche both *says and unsays* (Blondel 1991, 30–33, 36, 248, 258) his principle of a freely composing force.

As early as "On Truth and Lying in a Non-Moral Sense," Nietzsche suggests that the world beyond the body is a complex of forces. According to perspectivalism, these can constitute an infinite array of forms: ". . . and the whole of nature cavorts around men as if it were just a masquerade of the gods who are merely having fun by deceiving men in every shape and form" (TL 151); "All events, all motion, all becoming, as a determination of degrees and relations of force, as a *struggle*" (WP 552/1887)[23]; ". . . reality: change, becoming, multiplicity, opposition, contradiction, war" (WP 584/1888). For any subject however, the external world forces as such are absolutely inaccessible. The individual (whether animal, plant, or mineral) *interprets* according to its relation to the object forces. Later works such as *Beyond Good and Evil*, *On the Genealogy of Morals*, and some notes of *The Will to Power* further develop Nietzsche's anti-metaphysics perspectivalism by renouncing the subject altogether. "At last, the 'thing-in-itself' also disappears, because this is fundamentally the conception of a 'subject-in-itself.' But we have grasped that the subject is a fiction. The antithesis 'thing-in-itself' and 'appearance' is untenable" (WP 552/1887). Nietzsche's reasons for invoking a perspectivalism that challenges a metaphysical foundation are already in place in "On Truth and Lying in a Non-Moral Sense." Here, the "thing-in-itself" has disappeared. Nonetheless, Nietzsche in "On Truth and Lying in a Non-Moral Sense" continues to speak of the aesthetically creative subject. He recognizes that human perspective operates according to certain common necessities. Although two humans perceiving the same object from the same vantage point do not have to perceive it the same way, they cannot change their perception at will.

> but when that same image has been produced millions of times and has been passed down through many generations of humanity, indeed eventually appears in the whole of humanity as a consequence of the same occasion, it finally acquires the same significance for all human beings, as if it were the only necessary image and as if that relation of

> the original nervous stimulus to the image produced were a relation of strict causality. (TL 149)

That a picture is not a necessary correlate of a particular nerve-stimulus does not mean that human beings have the ability to change or shape the picture at will. "Man is necessity down to his last fibre, and totally 'unfree,' that is if one means by freedom the foolish demand to be able to change one's *essentia* arbitrarily, like a garment" (PTG 7). This marks a demand that "every serious philosophy has rejected with proper scorn" (PTG 7).

A variation of pictures might arise from multiple instances of the same nerve-stimulus. "Even the relation of a nervous stimulus to the image produced thereby is inherently not a necessary relationship" (TL 149). But the differentiation of pictures does not happen at will. Nietzsche's later work more clearly develops a sense of fluidity of forces constituting the relation between the human body and human reflection. Chapters 2 through 5, for instance, illustrated that in Nietzsche's *On the Genealogy of Morals*, human beings appear as functions of changing biological and socio-cultural forces in a manner that defies stable species definition.

As for the example of a picture arising from a nerve-stimulus, this displays metaphor in two ways. First, it does not duplicate essentially or literally that to which it corresponds. Second, just as a number of different metaphors might be used to illustrate a single topic, a number of possible pictures could delineate a nerve-stimulus, not arbitrarily or at will, but according to variable circumstances that affect the body unconsciously. It is for this reason that Nietzsche distinguishes the daily life of an early Greek from that of a modern-day scientist. The former he says is more similar to a dream than to the habitudes of the thinker "sobered by science" (TL 151). In a dream one believes that at any instant, all things are possible. Such a belief, suggests Nietzsche, characterized the attitude of the early Greeks during waking experience: "If, one day, any tree may speak as a nymph, or if a god can carry off virgins in the guise of a bull, if the goddess Athene herself is suddenly seen riding on a beautiful chariot in the company of Pisistratus through the market-places of Athens—and that was what the honest Athenian believed—then anything is possible at any time, as it is in a dream . . ." (TL 151). Whether Athenians actually did see Athena and her team in the market is not the issue. What is at stake for Nietzsche with respect to the early Greeks is their belief that their daily experience does not delimit reality per se. In other words, they might believe that a dynamic woman in the agora could be Athene transposed or disguised. If waking life is governed by certain laws of human subjectivity like space and time, the appearances produced by these might be understood as transpositions of a fantasy world that exceeds the limited appearances of human subjectivity. Time and space "we produce within ourselves and from ourselves with the same necessity as a spider spins" (TL 150). Because one's view is limited by subjective forms, the external world of forces is utterly unknown to us except as an anthropomorphic valuation—a metaphor.

If the human subject cannot fathom the world beyond, then central to Nietzsche's saying regarding a freely creating force is an unsaying. We can see this in two senses. First, we see this in his habit of making a statement and recanting it (Blondel 1991, 33). For instance, he says that our metaphors (i.e., our metaphor for the external world as a freely creating force) do not correspond to the essences of things: "and yet we possess only metaphors of things which in no way correspond to the original entities" (TL 144). He then *unsays* this: "For the opposition we make between individual and species is also anthropomorphic and does not stem from the essence of things, although we equally do not dare to say that it does *not* correspond to the essence of things, since that would be a dogmatic assertion and, as such, just as incapable of being proved as its opposite" (TL 145). Second, Nietzsche both says and unsays his principle regarding a freely composing force (hence I call it a [non]foundational foundation), in the two-fold manner in which he describes deception. First, there is the deception constitutive of his idea of aesthetic way of relating. This is based upon the notion that as subjects, our perception is not "correct" in any essential or absolute sense but is instead shaped by the forms of space and time and the subsequent sums of relations that follow from these and constitute our subjectivity. Thus, there is the basic deception brought about by the inevitable carrying-over constitutive of subjectivity.

Second, deception occurs regarding a human being's impulse to truth. If our bodily and intellectual experience is constituted by metaphors that are in themselves radical translations of object forces from one sphere over into another, and transpositions at least involve changing the object if not distorting it, then how is it that human beings have a concept of truth? Our will to truth, suggests Nietzsche in this early text, points to our denial of the greatest "fact" of human experience: that we are first and foremost self-deceiving (TL 143). "Only through forgetfulness could human beings ever entertain the illusion that they possess truth to the degree described above" (TL 143).

If deception takes place in this dual manner, then such deception also applies to Nietzsche's saying regarding a freely composing force. In other words, this assertion, like all assertions is necessarily distorted. The self-deceptive aspect of human being undermines every saying. Because Nietzsche develops this view of deception he also commits himself to developing a *style* of thought that says and unsays. Interestingly, to unsay his suggested metaphysical limit—the freely creating force—is to unsay the ground of the two-fold deception. That is, to unsay the principle that demands this unsaying. By characterizing human subjectivity in terms of deception, Nietzsche's assertion when applied to itself is also put in question. This movement of saying and unsaying occurs in the body of Nietzsche's text as early as "On Truth and Lying in a Non-Moral Sense." It acts as a structural metaphor for the so-called self, unconscious body, and external world about which we can say almost nothing for certain. Nietzsche speaks of this opaqueness continually after 1872.[24]

As early as "On Truth and Lying in a Non-moral Sense," human experience as artistically creative subjects, involves metaphor. Moreover, these metaphors are strung together by a meta-metaphor of saying and unsaying. And so, after *The Birth of Tragedy*, Nietzsche's faith in music as the best metaphor for a world essence drops away. Although metaphor in terms of translation, carrying-over, metamorphosis, and self-dispossession still plays the pivotal role constituting human life, Nietzsche recognizes that life as and through metaphor tells us nothing for certain about the 'external' world, except that it appears. Because our experience is constituted by such metaphor—the inessential and improper—Nietzsche creates a devaluation of the intellect. Human perspective as metaphor and through metaphor appears to be first and foremost deceptive.

Metaphor and the Body in Nietzsche's Writing, 1873–1888

Nietzsche's devaluation of the intellect apparent in "On Truth and Lying in a Non-Moral Sense" continues in Nietzsche's subsequent works. Moreover, in his later work, Nietzsche treats the body as primary. One might be led to believe that the primacy Nietzsche assigns to the body makes bodily sensation and instinct more "real" than conscious idea. If the bodily metaphor lies closer in proximity to the external world would not its translation be more reliable than subsequent translations? For instance, would not an idea of water be less real than the sight or touch of it? Eric Blondel notes that Nietzsche rejects such reasoning in that it would reduce reality to mechanistic materialism and inevitably draw him back to a metaphysical dualism.[25] Nietzsche clearly expresses his intent to avoid such a result: "Neither of the two explanations of organic life has yet succeeded: neither the one that proceeds from mechanics *nor the one* that proceeds from the spirit. I stress *this last point*. The split is more superficial than we think. The organism is governed in such a way that the mechanical world, *as well as* the spiritual world, can provide only a symbolical explanation."[26] The concept of the "symbolical" implies a subordinating of the concept to the metaphor. Through his metaphors Nietzsche tries to counter both a reduction to materialism and to spiritual idealism. In the former the body is objectified and becomes supposedly real, whereas in the latter corporeality dissolves into idea, which itself becomes the supposed real. According to Blondel (1991, 204–205), Nietzsche challenges both outcomes by attempting to make the body primary and still avoid an empiricist position. This bars one from viewing ideals as lower orders of reality derivative of the body or from falling into a dualism that is a simple inversion of Platonic dualism. Instead of a mechanistic reductionism or metaphysical dualism, Nietzsche discloses a dynamic non-dualism, based on "a union and separation which exists between the ideal and the body" (205). In order to block a reductive materialism such a non-dualism will not be based on the dualistic branching out of extension and thought into attributes from a single substance.[27] Instead it will be constituted by interpretation "spelt out as metaphor and through metaphor" (205).

I have already shown that for Nietzsche, all concepts are at their basis metaphors. In showing this, Nietzsche not only subordinates the concept to the metaphor, but also subverts a strict mind-body distinction. Concepts are coarse assimilations of the body's manifold signs (BGE 230). They are not forms that exist prior to or separate from physical valuations. If Nietzsche views metaphor as central to the human body and intellect, one sees why an appropriate mode for *writing* about life would be to write metaphorically. By employing a writing style that uses metaphors of physiology for describing the intellect and metaphors of cognition for describing the body, Nietzsche further challenges a Cartesian dualism. The one (e.g., the mind) becomes a metaphor for the other (the body) and vice versa: Our affects are like souls: they think; Our intellect is like a stomach: it digests. The following excerpts display bodily drives as conscious, thoughtful and judicial:

> Movement is symbolism for the eye; it indicates that something has been felt, willed, thought. (WP 492/1885)

> —pleasure is only a symptom of the feeling of power attained, a consciousness of a difference. (WP 688/1888)

> We find that the strongest and most constantly employed faculty at all stages of life is thought—even in every act of perceiving and apparent passivity! Evidently, it thus becomes most powerful and demanding, and in the long run it tyrannizes over all other forces. Finally it becomes "passion-in-itself". (WP 611/1883–1888)

> Pain [is] intellectual and dependent upon the judgement "harmful": projected. (WP 490/1885)

If, in the above excerpts, Nietzsche attributes to the body qualities traditionally linked with mind, in the following he does a similar reversal, but this time with respect to mind. He describes intellect, spirit, and consciousness,[28] and ideas associated with these—learning and virtue—in physiological terms:

> Learning changes us; it does what all nourishment does which also does not merely "preserve"—as physiologists know. (BGE 231)

> By applying the knife vivisectionally to the chest of the very *virtues of their time*, they [philosophers] betrayed what was their own secret: to know of a new greatness of man, of a new untrodden way to his enhancement. (BGE 212)

> The spirit's power to appropriate the foreign stands revealed in its inclination to assimilate the new to the old, to simplify the manifold. . . . Its intent in all this is to incorporate (Einverleibung) new "experiences," to file new things in old files–. . . all of which is necessary in proportion to a spirit's power to appropriate, its "digestive capacity," to speak metaphorically–and actually "the spirit" is relatively most similar to a stomach. (BGE 230)

The body interprets, evaluates, and thinks. The intellect tastes, digests, and eliminates (BGE 231; WP 229/1888). In describing certain shared characteristics, Nietzsche transposes the body and the intellect with respect to their attributes, thereby suggesting a loss of determination for both. By characterizing mind more by that which has traditionally been considered physiology and physiology more by attributes traditionally associated with mind, Nietzsche retains a soft opposition. These characterizations and opposition are meant both literally and nonliterally. Nietzsche's aim seems to be to problematize a clear line of demarcation between the mental and the physical. By avoiding a merely literal reading of his respective mind and body attribute reversals, Nietzsche avoids reconstructing the very dualism his figurative inversion calls into question.

It might seem that if Nietzsche attempts to avoid physical reductivism as much as spiritual reductivism by depicting both the body and intellect through metaphors of the other, that he would value both the body and intellect equally. He does not. We have seen *The Birth of Tragedy* and "On Music and Words" prefigure this move by robbing the human subject of contact with anything but the inessential and non-in-itself: metaphor. Concepts of the intellect turn out to be metaphors. "On Truth and Lying in a Non-Moral Sense" goes a step further in characterizing the power of cognition as the power of dissimulation. In Nietzsche's writings after 1873 this devaluation of the intellect develops into a primacy of the body over the spirit. One can see such a primacy and concept of body exposed in excerpts below from *The Will to Power*, *The Gay Science*, and *Beyond Good and Evil*, and spanning 1882–1888.

It is essential "to start from the *body* and employ it as guide. It is the much richer phenomenon" (WP 532/1885). "Belief in the body is more fundamental than belief in the soul: the latter arose from unscientific reflection on the body" (WP 491/188b5–6). "All perfect acts are unconscious and no longer subject to will . . . a degree of consciousness makes perfection impossible" (WP 289/1888).

"In fact, this unconsciousness belongs to any kind of perfection: even the mathematician employs his combinations unconsciously" (WP 430/1888). "Genius resides in instinct; goodness likewise" (WP 243/1887–1888).

The primacy of corporeity expressed above is also conveyed in Nietzsche's critique of "Socratism," which accuses Euripides, Socrates, and the philosophical thought that succeeds them of reversing the order of body and mind.[29] They "severed the instincts from the *polis*, from contest, from military efficiency, from art and beauty" (WP 435/1888). In the Classical Greek philosophers Nietzsche sees "a decline of the instincts: otherwise they could not have blundered so far as to posit the *conscious* state as *more valuable*." The Greeks failed to understand that the "intensity of consciousness stands in inverse ratio to ease and speed of cerebral transmission" (WP 439/1888). By "positing proofs as the presupposition for personal excellence" the Greeks tore moral judgments "from their conditionality, in which they have grown and alone possess any meaning, from their . . . Greek-political ground and soil, to be denaturalized under the pretense of sublimation" (WP 430/1888).

To counter the entrenched dogma privileging spirit over matter implicit in a Socratism still widespread, Nietzsche elevates the body and denigrates conscious intellect. He describes the latter as a mere appendage of the former. "It [consciousness] is not a directing agent but an organ of the directing agent" (WP 524/1888). Body is more fundamental than the soul, "the latter arose from unscientific reflection on the body" (WP 491/1885–1886). "We shall be on our guard against explaining purposiveness in terms of spirit. . . . The nervous system has a much more extensive domain; the world of consciousness is added to it" (WP 526/1888). "Consciousness is the last and latest development of the organic and hence also what is most unfurnished and unstrong" (GS 11). Religion, psychology, metaphysics, wisdom, and rationalism: all "stand truth on its head" (WP 576/1883–1888; WP 608/1886–1887; WP 434/1888). We must "seek perfect life where it has become least conscious" (WP 439/1888).

Nietzsche's response to Socratism—his campaign to revalue for human beings the body and to return human beings to nature—is visible: "The basic text of *homo natura* must again be recognized." To do this we must "translate man back into nature" (BGE 230). "Error reached its peak when Schopenhauer taught: the only way to the 'true,' to knowledge lies precisely in getting free from affects, from will" (WP 612/1887). Humankind needs "[t]o win back for the man of knowledge the right to great affects" (WP 612/1887)!

Typically unmentioned by Nietzsche's commentators is the fact that Nietzsche's writings are certainly not the first to prioritize corporeality over intellect, nor to do so against a backdrop depicting human experience as creative artistry (Moore 2002, 86). As Moore points out, prior to the Kantian influence that would locate the supposed origin of aesthetic judgment transcendentally, several eighteenth-century British philosophers, such as Edmund Burke, Uvedale Price, and Daniel Webb, located it physiologically (86–87). Burke's physiologically rooted aesthetics as described by Moore, for example, associates "the apprehension of beauty" with a process of "multiplication of the species, producing 'the passion of love in the mind' and the accompanying pleasurable sensations of melting or languor by causing the fibres of the body to relax below their natural tone" (86).[30]

Within ten years of its publication in 1859, however, Darwin's *Origin of the Species* (1859/1985) would "become one of the dominant discourses of the latter half of the nineteenth century" (Moore 2002, 2). Its influence would encourage the work of certain scientists and philosophers to again position physiological processes as a basis of human experience.[31] If, in this light, Nietzsche appears unoriginal with respect to giving corporeality a sort of primacy, his brand of aesthetic physiology is unique in its general antipathy towards teleological overtones. Moore reminds us of Nietzsche's well-known assertion in *Beyond Good and Evil* (1966) that "there are no moral phenomena, instead only moral interpretations of phenomena" (KSA V, 92, qtd. in Moore 2002, 64–65). Even Darwin, well known for typically suspending judgment on the metaphysical significance of his evolutionism, generally succumbs to a form of teleology. As Moore (2002) points out, Darwin views survival as a "primary biological imperative" (66) and thus, as a sort of species end. It is this feature of an implied metaphysics of ends that for Moore, accounts for Nietzsche holding a variant of evolutionary theory that is nonetheless in Nietzsche's view "anti-Darwinian" (66, 110). Darwin's implied teleology is more subtle than that of someone like Herbert Spencer, whose explicit language of ends is more typical of Nietzsche's contemporaries. For Spencer (1879, 82) for example, Darwin's ideas provide material for an evolutionary ethics. Accordingly the gauge for "good" and "bad" becomes the usefulness of an item or behavior for maximizing pleasure—pleasure not only for the organism but also for the species (Moore 2002, 64). And so it is important to note that Nietzsche's uniqueness emerges not in his giving the body primacy, nor in his understanding of this primary body in more-than-literal terms—that is as creative artistry in the sense of a *Kunsttrieb* or *Metaphertrieb*, but in the way he describes the two former themes so as to undermine a metaphysics of ends.

If one examines Nietzsche's metaphors for physiology, for instance, one sees they are not graphic, concrete images. They demand that one creatively "translate" them (Blondel 1991, 214–220). This demand points as much to the results of such a transposition as to the process itself: *interpretation* (Blondel 1991, 214–220). The process of interpretation deflates, I shall show, both a spiritual and material teleology. We can view interpretation as *das Übertragen*: translation, carrying-over, transforming. Or, as characterized above in chapter 2, relation: relatives, metamorphoses, putting-to-use, turning towards new ends, and exchanging. Moreover, the metaphors depicting the body almost always imply *evaluation, commentary, thought,* and *perspective* (Blondel 1991, 214–220). Thus, not only does the metaphor as a modality link the body to interpretation, but so does the subject-matter of the metaphors themselves. Nietzsche makes the body primary and he attempts to avoid reductivism via a metaphorical language that points not to the body as purely mechanistic or spiritual but to the body as an interpretive field both uniting and separating these modes (Blondel 1991, 205). In this manner metaphors articulate not only "the *style* of *Nietzsche's* descriptive metalanguage" but also "the interpretive status *of his object* itself: the body" (Blondel 1991, 205). If his object—the body—is also the site in which human experience of bodily sensations and consciousness appear to be (non)foundationally rooted, it seems that metaphors say more than simply the style of Nietzsche's metalanguage: they constitute its source and formation too.

In Nietzsche's 1873–1888 writings, we can envision metaphor as the source of Nietzsche's meta-language—which is to say, as the body and its interpretive status—and we can do so in two ways. The first can be described as a continuum of metaphors. By continuum of metaphors, I mean a dynamic constellation of forces translating and transposing myriad constitutions of forces between the world, bodily drives and intellect. These translations are metaphors in the sense described in "On Truth and Lying in a Non-Moral Sense." They are "an allusive transference" of forces from one sphere to another (TL 148).

The world as chaos can be said to represent one extreme of the continuum. Nietzsche does not propose this chaos in any but an hypothetical sense. This hypothesis itself is a metaphor produced by the "other" extreme of the continuum, the so-called intellect. Whereas the world multiplies force, the intellect simplifies this chaos. Occupying an intermediary place between these, bodily instincts and affects translate and evaluate the chaos of the world. Thus, in comparison relative to the hypothesized absolute world multiplicity the instincts carry out a sort of simplification of data.

> One should not understand this compulsion to construct concepts, species, forms, purposes ("a world of identical cases") as if they enabled us to fix the real world; but as a compulsion to arrange a world for ourselves in which our existence is made possible—we thereby create a world which is calculable, simplified, comprehensible, etc., for us.
>
> This same compulsion exists in the sense activities that support reason—by simplification, coarsening, emphasizing and elaborating, upon which all "recognition," all ability to make oneself intelligible rests. (WP 521/1887)

But if in the transcription the instincts are forced to simplify, they do so only in order to pluralize again. That is, rather than institute order and organization among the drives, they encourage change and pluralization. Everything that enters consciousness as an apparent unity comes from an inner instinctual world teeming with forces at cross-purposes, in "chaos" and in "confusion" (WP 594/1883–1888).

> Science—this has been hitherto a way of putting an end to the complete confusion in which things exist, by hypotheses that "explain" everything—so it has come from the intellect's dislike of chaos.—This same dislike seizes me when I consider myself: I should like to form an image of the inner world, too, by means of some schema, and thus triumph over intellectual confusion. Morality has been a simplification of this kind: it taught that men were known, familiar.—Now we have destroyed morality—we have again become completely obscure to ourselves! I know that I know nothing about myself! (WP 594/1883–1888)

In relation to the intellect, the inner world of bodily instincts could be said to be, for Nietzsche, a streaming plurality. The intellect, which Nietzsche also refers to as "consciousness," and "spirit" takes the more complex and simplifies it: "That commanding something which the people call "the spirit" wants to be master in and around its own house and wants to feel that it is master; it has the will from multiplicity to simplicity, a will that ties up, tames and is domineering and truly masterful" (BGE 230).

The transposition from world to body to intellect involves a process of increasing simplification. If transposition only moved in the direction from the multiplicitous to the simple it would seem that the simplicity achieved by conscious intellection would be a goal, a teleological end. But Nietzsche also discloses this movement headed in reverse. Indeed, consistent with his strategy of prioritizing the body, Nietzsche's language valorizes and even encourages the "return" of we "men of knowledge" to the body (GM preface 1). Nietzsche speaks of the body "incorporating" (absorbing and transforming) into its multiplicity the more reductive metaphors of consciousness. "To this day," Nietzsche writes, "the task of *incorporating* [*sich einzuverleiben*] knowledge and making it instinctive is only beginning to dawn on the human eye and is not yet clearly discernable" (GS 11). The body to which we would "return," is not any we recognize from a perceivable past. It is no identifiable mass or mechanism or physiology but appears as a complex blend that interpenetrates psyche and body and defies the ideas of our inherited dualisms. We see this conflation in Nietzsche's description of forgetting as "'inpsychation.'" To digest our experience by forgetting it, writes Nietzsche, a process that "one might call . . . 'inpsychation,'" is an "active and in the strictest sense positive faculty of repression" (GM 2, 1).

Like "the thousandfold process, involved in physical nourishment—so called 'incorporation,'" the forgotten "inpsychated" experiences likewise enter "our consciousness as little while we are digesting it" (GM 2, 1). Forgetting, therefore, provides another example of transposition in the direction of the manifold "physiological." We see, then, that transposition moves toward both the more simple (intellection) and the more complex (incorporation), undermining any naturally ideal or material teleology.

Nietzsche's continuum of metaphors, if valorizing a nonteleological transposition that "returns" us toward the so-called "multiple," and "natural," also discloses and sustains the supposed "simple," and "supernatural." This dual recognition is visible in *The Antichrist*. There Nietzsche critiques a certain Christian tradition wherein "'spirit'" and "'divinity'" have become ideals of human dignity and valor. "We no longer trace the origin of man in the 'spirit', in the 'divinity,' we have placed him back among the animals. We consider him the strongest animal because he is the most cunning: his spirituality is a consequence of this" (A 14). Nietzsche places persons "back among the animals" while distinguishing them as the "strongest animal." Human beings' "cunning" sets them apart and "spirituality" arises from the latter. Nietzsche implies a continuum of metaphors according to which we both return to our animality and exist with an experience of "'spirit.'" Nietzsche's continuum of metaphors exposes the spiritual phenomena and the body as one unit, but this is a plural unity.[32] The metaphor of the body serves as a principle of plurality; metaphors such as the ideas of will, spirit, consciousness, and free-will—ideas that pertain to a human being's spiritual tendency—provide a principle of unity.

If spirit and the body are a plural unity, what is the significance of understanding the body itself as primary? In giving the body an interpretive status, spirit itself seems slyly reintroduced into the body. In what sense does the body's primacy have any significance if the body may in a sense be so-called "spirit." What is the significance of understanding the body both as fundamental and as interpretation (Blondel 1991, 201)?

To articulate this significance, let us note that for Nietzsche the body is characterized by instincts (*Instinkte*) or drives (*Triebe*) that "constitute reality as they interpret it" (Blondel 1991, 206). By viewing the body as *interposed*[33] between the external world and the conscious intellect, Nietzsche invites two consequences. First, the moral subject[34] and the concept are replaced, respectively, by a disintegrated subject and "concept" understood *as* metaphor. What is important is that the concepts of the moral subject and of the concept are interpretations (or relations in the sense of relatives, metamorphoses, putting-to-use, turning towards new ends, and exchanging, as described in chapter 2) based on the bodily instincts. Second, "intellect becomes the *instrument* of an unconscious interpreting body."[35] The result, as Blondel points out, is not simply that a traditional Christo-Platonic dualism is overturned (1991, 206). There is indeed a "change of order," but more noteworthy: the body *as interpretation, as relation* shows itself as fundamental (Blondel 1991, 206). This explains why Nietzsche's metaphors expressing the primary status of the body do not ground the body but point to the process of interpretation itself. By understanding the body as fundamental interpretation as and through metaphor we see that human beings and culture are *more* than the body, and yet, all that they are can be said to begin through the body.[36]

The second way that metaphor constitutes the status of the body as interpretation is through a "metaphor of foundation" (Blondel 1991, 209). Despite above suggestions to the contrary, Nietzsche's assigning primacy to the body as fundamental interpretation, nevertheless, brings the concept of body near a foundational status. Indeed, such primacy leads Blondel to call the concept of body, for Nietzsche, a "metaphor of foundation" (209). This quasi-foundational appearance of the concept of body could lead some to believe that the metaphorical status in the case of the concept of foundation actually inverts itself, and that indeed, the body is "foundational" (209). As it turns out, Nietzsche places the two concepts of will and soul—often viewed in the Western tradition as grounding principles—"underground" in the enigmatic depths of our body: "In this way the person exercising volition adds the feelings of delight of his successful executive instruments, the useful "under-wills" or under-souls—indeed, our body is but a social structure composed of many souls" (BGE 19). Nietzsche speaks of the body as a world of "under-wills" and "under-souls" (BGE 19) and of an "underworld of organs" (GM 2, 1). He continually reminds us to mistrust consciousness. "The nervous system has a much more extensive domain; the world of consciousness is added to it" (WP 526/1888). Because consciousness "is not a directing agent but an organ of the directing agent [body], "we are to turn to the body. It is essential "to start from the *body* and employ it as guide. It is the much richer phenomenon" (WP 532/1885).

Given such founding descriptions of the body, one might be tempted, notes Blondel, to compare this metaphor of foundation to a Marxian image of foundations that inverts Platonic dualism. In such a case the ground is the material infrastructure [*Unterbau*] and consciousness is merely its reflection (Blondel 1991, 204). For Nietzsche, however, no such inversion takes place. Although he calls upon humans to return to the body, to "translate man back into nature" out of the spirit's "moral word tinsels" and "verbal pomp," he makes no illusions about just what we are returning to (BGE 230). The underworld proves just as opaque as the gloss of superficial consciousness. "Actions are *never* what they appear to us to be! We have expended so much labour on learning that external things are not as they appear to us to be—very well! the case is the same with the inner world" (D 116). The notion of underworld as metaphor for the body points less to the body as a hidden "ground," than to the body as a philological labyrinth of plurality (Blondel 1991, 205). Rather than a philosopher interested in foundational causes, Nietzsche's investigations into the body, suggests Blondel, show him to be an intensely skeptical interpreter of meaning. He ferrets out meaning's undecipherability. "However far a man may go in self-knowledge, nothing however can be more incomplete than his image of the totality of *drives* which constitute his being. He can scarcely name even the cruder ones: their number and strength, their ebb and flood, their play and counterplay among one another, and above all the laws of their *nutriment* remain wholly unknown to him" (D 119). In his writing through and about the body, Nietzsche discloses an interpretive "axis" of thought (Blondel 1991, 207). This axis mixes but does not identify physiology and spirituality.[37]

It is intellect, however, and not the bodily affects, which fails to express the complexity and difficulty of understanding meaning. As early as 1873 Nietzsche indicates that not only is human access to a foundational "cause" or "world essence" unfathomable, but that the human intellect cannot understand the complex ways of the body (TL 142, 145–147). Via our conscious intellect, that is, our "ability to sublimate sensuous metaphors into a schema . . . [we] dissolve an image [*Bild*] into a concept [*Begriff*]" (TL 146). But spirit's conscious ideas and "linguistic means of expression are useless for expressing 'becoming'" (WP 715/ 1887–1888). For expressing "reality: change, becoming, multiplicity, opposition, contradiction, war" in other words, for expressing the complexities of meaning, it is useless (WP 584/1888).

The spirit can *simplify* the body's dark world of plurality but it cannot comprehend or read it: "That commanding something which the people call "the spirit" wants to be master in and around its own house and wants to feel that it is master; it has the will from multiplicity to simplicity, a will that ties up, tames, and is domineering and truly masterful" (BGE 230). Although the intellect can subsume and master, "there is no ground whatever for ascribing to spirit the properties of organization and systemization" (WP 526/1888). Instead, it offers "the most superficial, most simplified thinking" (WP 527/ 1886–1887). "Thoughts are shadows of our feelings—always darker, emptier and simpler" (GS 179). Nietzsche's human-herd members, those "*levelers*" [Nietzsche's emphasis] who set "truth happily on her head" as they cry out for "'equality of rights,'" could be a metaphor for spirit (BGE 44). Rather than grasp the multiple differences of the body's underworld, they seem, like spirit, Nietzsche suggests, to flatten them.

Nietzsche investigates intellectual thinking, and finds it incapable of grasping the intricacies of meaning expressed through the body. If spirit—which remains on the surface—fails to grasp the complex underworld, is body, with its descending depths, able to do so? What would it mean to effectively grasp an object and what would one grasp were one to do so? We know that Nietzsche's 1873 view of the human subject and aesthetic relation already contests the notion of an in-itself or essential world truth. There appears to be nothing absolute to grasp. If grasping or beholding the so-called real is understood as that which the body constitutes as it interprets, then Nietzsche, it would seem, views so-called reality in terms of the body *as* interpretation: the body constitutes reality as it interprets it.

Nietzsche explores this reality by exploring the body, that is, by going beneath the surface. As a "subterranean man" and "cave hermit" who "tunnels and mines and undermines" (D 1) he interrogates the ground of things from beneath it (Blondel 1991, 213). Thus, Nietzsche's foundational metaphor for the primacy of the body also invokes the metaphor of depth (Blondel 1991, 213). Underneath each ground is another to stand on, behind each cave, is another opening deeper back.

> Indeed [the hermit] will doubt whether a philosopher could *possibly* have "ultimate and real" opinions, whether behind every one of his caves there is not, must not be, another deeper cave—a more comprehensive, stranger, richer world beyond the surface, an abysmally deep ground behind every ground, under every attempt to furnish "grounds." Every philosophy is a foreground philosophy—that is a hermit's judgement: "There is something arbitrary in his stopping *here* to look back and look around, in his not digging deeper *here* but laying his spade aside; there is also something suspicious about it." Every philosophy also *conceals* a philosophy; every opinion is also a hideout, every word also a mask. (BGE 289)

The association of the metaphor of foundation with the metaphor of depth shows that rather than a single foundation, there are multiple ones. That each (non)foundation may present itself as a ground, depends upon the perspective of the philosopher. The metaphor of depth, however, serves to *transpose* the traditional concept of foundation into a perspectivalism. One particular cave as opposed to another may appear, depending upon one's perspective, to be the final and ultimate cave. One who is like the hermit, who does not forget the metaphor of depth, will "see through" the appearances of foundation. He or she will remember that each foundation also conceals a foundation; each "philosophy also *conceals* a philosophy" (BGE 289).

One might think that in taking his survey of meaning underground, Nietzsche would be able to contrast his findings with the aetherial *surfaces* of conscious spirit (Blondel 1991, 213–214). I have shown earlier that Nietzsche does not reduce the intellectual ideal to the bodily affect or vice versa, but that he also avoids severing the two. In what sense is the underworld of instinctual drives like the conscious world of mind? In what way is it different?

We have said that consciousness stays on the surface. We can also assert that consciousness creates surfaces: "We must show to what extent everything that is conscious remains on the surface; to what extent action and the image of action are different, how *little* we know of what precedes an action . . . to what extent thoughts are merely images, to what extent words are merely [*nur*] signs [*Zeichen*] of thoughts, the impenetrable nature of any action (KSA 10, 654). A word, a thought, or an image is merely a transposition of another constitution of forces that is itself a transposition and so forth. Nietzsche not only invokes consciousness here but also embodiment. The concept of action can be said to refer to the so-called physical and the term "action" in the citation above suggests a metaphor along a nondualistic physiology-intellect continuum resting at the physiological extreme. The metaphors referred to imply transposition along the continuum increasingly in the direction of the so-called conscious: action → an image of an action → a thought as an image → a word as a sign for a thought. The action, however, also constitutes a surface. It covers-over and thereby hides "what precedes an action" (KSA 10, 654). Thus, not only intellectual but also physiological representations can appear as surfaces. Moreover, Nietzsche describes the bodily drives as "thinking." "Man, like every living being, thinks continually without knowing it; the thinking that rises to *consciousness* is only the smallest part of all this—the most superficial and worst part" (GS 354). The intellect, then, is not the only producer of thought, surfaces and signs, but so is the body. "All of thought comes to us as a surface and biased, so too our desires" (KSA 7, 446). Both an instinctive and conscious thought cover over a set of plural forces to which each as a "surface" refers. In this way we can understand both the affect and the idea as signs. Furthermore, each sign as the site of the indiscernible plurality it signifies in spite of itself, constitutes a plural unit (Blondel 1991, 214).

"Order," "relation," and "text" only come into existence through the initial reductive capacity of our bodily drives. Yet, this ability to reduce differs from that of our intellect. The body's drives simplify the world chaos in order to pluralize it; the intellect's ideas simplify the bodily affects in order to unify them. The affects remain "surfaces," however, in two senses: they transpose their objects—one another and the absolute multiplicity of the world; and they are inevitably reductions of world multiplicity to which they continue to refer by presupposing it.

Thus, both the intellect and the body create signs, that is, *surfaces* rather than *substances*. A surface, unlike a substance in the proper sense, is always under-stood by or under-stands something else. It transposes something "below" or "above" it. Thus, the metaphor of depth for the bodily underworld is not "depth" as opposed to "surface," but depth constituted by surfaces—an infinite layering of signs (Blondel 1991, 214). The body's opaque depth does not indicate the hiding of a foundation but the hindrances barring one from understanding meaning.

PART IV. CONCLUSION

It is now possible to see the body as the site that introduces text into the world, moreover, as a field that generates metaphors while interpreting them. In this manner, and with insights from Blondel, Nietzsche's writing can be seen showing the body to be not only the generative source of his text, but the text's primary object. One might say that Nietzsche's corpus is a writing of the body, but as such, it not only displays the body but also life and spirit.

We are also in a position to answer the question posed at the end of chapter 5 that initiated the inquiry here. We asked there whether Nietzsche would say that meaning really produces itself in the non-dual, mobile, reciprocally shaping manner the concepts of dynamic non-dualism and relation display. In this chapter we saw that as early as 1873, that Nietzsche does away with a presupposition of primal cause and develops a notion of human experience constituted as and through metaphors transposing one another along a world-body-intellect continuum that invites transposition bidirectionally. I showed that Nietzsche's metaphors for the body and intellect describe each in terms of the other, undermining a metaphysical dualism. These metaphors point not to the intellect but to the body as primary.

What is one to make of this primacy? In examining the bodily metaphors one finds that they are not transparent. They demand that one creatively translate them. As such, the metaphors describing the body point as much to the meanings of such metaphors as to the process itself: interpretation. What is interpretation? It is *das Übertragen*: translation, carrying-over, transforming. In chapter 2 this was characterized as *relation*: relatives, metamorphoses, putting-to-use, turning towards new ends, and exchanging. Moreover, the symbolic metaphors depicting the body almost always imply evaluation, commentary, and thought. Thus, not only does the metaphor as a modality link the body to interpretation, but so do the themes of the metaphors themselves. As Blondel suggests, Nietzsche makes the body primary and attempts to avoid reduction.

Nietzsche does this via a metaphorical language that points not to the body as purely mechanistic or spiritualistic, but to the body as an interpretive field both uniting and separating these modes (Blondel 1991, 206–207). In this manner metaphors articulate not only the meta-language of Nietzsche's writing style but also the interpretive status of life's (non)foundational primacy: the body (205). When Nietzsche goes underground to investigate the body and reveals its perplexing multiplicity and depth of surfaces, he indicates the obstacles keeping one from a reliable hermeneutics of meaning

Nietzsche's assessment of intellectual thought shows it incapable of grasping the intricacies of meaning expressed and interpreted through the body. This led me to agree with Blondel that for Nietzsche, the bodily drives moreso than the intellectual ones, constitute so-called reality as they interpret it. For some this might suggest that "the real," for Nietzsche, is interpretation or relation. Yet on Nietzsche's own terms, reality as relation recoils in at least two ways. Indeed, relation is characterized by various kinds of interrelation, overturning, and exchange. This means that as relation "reality" is centrally characterized by a mechanism to overturn itself qua relation. In this sense, while Nietzsche might say that at any given time in Western history—so-called "reality" has been characterized by relation, he would recognize the possibility in some distant past or future of an overturning of such a supposed reality.

Second, and perhaps more important, Nietzsche's own view expressed in "On Truth and Lying in a Non-Moral Sense," and implicit in Nietzsche's subsequent works, suggests that human experience is first and foremost deceptive. Although this deception is not of the sort to make Nietzsche question the existence of the external world, it does make him question any saying that characterizes that world. If Nietzsche were to say that reality *is* something, he would most likely affirm that it is relation.

We saw in this chapter, nevertheless, that relation presupposes both change and self-deception. This suggests that as soon as Nietzsche would make the assertion that reality is relation, he would acknowledge the factor that would lead him to recant it. Because human experience is deceptive for Nietzsche, even the notion of human experience shaped as and through metaphors, as and through relations is subject to question. If metaphor is the meta-language of Nietzsche's writing style and points to the interpretive status of the body, we might say that the dialectical movement of saying and unsaying is the outermost limit of this meta-language. Although this limit prevents Nietzsche from saying something, it also prevents him from saying nothing. It directs us to a motion that posits and de-posits, creates, and destroys, a motion that may eventually subject saying and unsaying itself as a metaphysical limit[38] to its own movement.[39]

Whether such an overturning will happen in a way that reflects Nietzsche's story remains to be seen, in the meantime, however, we have revealed Nietzsche's view of the body as primary and as interpretation, a view that prefigures aspects of the phenomenology of Maurice Merleau-Ponty.

CHAPTER 7

Nietzsche after Nietzsche

Trauma, Language, and the Writings of Merleau-Ponty

Nietzsche's suspicion of "official" stories invites comparison with twentieth- and twenty-first-century tales of phenomenologists. Phenomenologists, of course, generally discard judgments about a possible "correct" story and replace them with analytical stories about the ways beings and being appear. Such a replacement renders expectations for possible official stories about beings and being virtually obsolete. Thus, it makes sense that if we extend Nietzsche's story toward tales told after his, that we extend it toward phenomenological ones. This extension will focus on one phenomenologist in particular, Maurice Merleau-Ponty (1908–1961). Merleau-Ponty, moreso than other phenomenologists, is known for emphasizing and privileging, as does Nietzsche, the role of the body in human experience.

PART I. CHAPTER OVERVIEW

At a concert recently, my friend Rita stood for the entire second half due to a cramp that stretched from her thigh to her lower torso. During the intermission, Donna, a physical therapist, cautioned Rita to stop her regular exercise routine for a couple of weeks. "Otherwise," Donna explained, "the associated neurons and muscles will learn to remember for good the painful and unhealthy way they are coordinating now." Many of us living in industrialized societies of the twenty-first century find ourselves, like Donna, speaking in surprising ways about the relationship between human learning, memory, and human anatomy. Just what do we mean when we say, "muscles learn and remember?" One deep-seated conviction in the West influenced by the legacy of Cartesian mind-body dualism is that thought (traditionally including language, memory and learning) happens above the neck, and whether or not "in" the brain (a matter of debate)—at least not *below* it.

Donna's statement to Rita, interestingly, harbors an implicit challenge to the Cartesian legacy: neurons and muscles below the cerebral cortex can learn and remember. Indeed, recent anatomy-adaptation research supports this view (Patterson 2001, 79; Wernig, Nanassy, & Müller 2001, 225–240; Wolpaw 2001, 119). The question of mind-body relation, therefore, is a question about boundaries—not merely about respective boundaries of mind and body but also about those of any item or being. Where does one "begin" and the other "end?" Do we know? And if we do not know, in what ways does a loss of belief in discrete entities or identities matter?

Even as physical therapists today invoke the language of muscle memory, for most of them (and us), our contemporary speech and perceptual structures keep our awareness bound to Cartesian concepts. Many of us may accept an analogy between muscular learning and memory and so-called ordinary human learning and memory, but will assume that these are categorically different—that the difference between human thought and muscle memory is more than a degree of complexity. Indeed, a prevailing unexamined assumption is that, at their root, human reflection and anatomy are fundamentally different. The influence of this view reflects how entrenched are the following three contemporary views that descend from modernity and mutually imply one another. These are the widespread conviction in Cartesian mind-body separateness (or dualism), belief in Enlightenment logic of Galileo-Newtonian mechanistic cause and effect, and commitment to the techno-scientific assumption of separateness of subjects and objects, selves and others, beings and things.

Through a reading of the writings of Merleau-Ponty, this chapter exposes as misguided such a Cartesian legacy through an examination of several contemporary practices and experiences. For Merelau-Ponty, animal corporeity[1] shows itself to be *communication*,[2] which varies from the primordial to the complex; such communication is rooted in *corporeity* that directs itself outward to delineate itself and make of itself a 'presentation.' Put differently, for Merleau-Ponty, corporeity exists as a self-showing that is self- and other-signifying, rooted in its own *desire*.[3] *Speech, I shall argue, when understood to be an expression of its origin—pure desire—shows itself to be less a mode of representation than a mode of bodily adaptation.*

That communication is rooted in corporeity proves important for several reasons. It shows the body as a whole displaying activity traditionally believed confined to the brain and to cognition. It implies a substitute for the traditional hierarchical chain of being—and its correspondent differences in ranks of beings:

a lateral plane of beings sharing equal rank. Finally, by showing that corporeity exists *as* communication, but that communication is *of* corporeity, my view is for an additional reason crucial. It shows as connected that which has often been seen as disconnected: idea and matter, creator and created, mind and body, psychology and biology.

Although many have critiqued mechanistic models of self and being entailed by a Cartesian mind-body dualism, and some using the philosophy of Merleau-Ponty, my examination is unique because it shows recent empirical research on physical and psychological trauma supporting the critique. Specifically, it analyzes lived experiences of traumatogenic narrative speech—in the aftermath of war or domestic violence—and recent research on adaptive physiology—in the aftermath of spinal cord injury.

Rarely have Merleau-Ponty's writings been placed in dialogue with contemporary literature of trauma studies.[4] Virtually none of the philosophical commentary on Merleau-Ponty acknowledges a concept of trauma implicit in Merleau-Ponty's work.[5] Nor does such Merleau-Ponty criticism recognize how a concept of trauma operative in current trauma research—that is, that of not only psychological but also physiological trauma—supplies new evidence for Merleau-Ponty's view of humans and nature, bodies and psyches, as interpenetrating.

PART II. LANGUAGE AS AND OF CORPOREALITY

All perceptual structure (e.g., that of space, time, sight, sound, smell, taste, touch, abstract reflection) can be said to begin for Merleau-Ponty, *as* and *of* corporeity. Moreover, corporeity exists as a structure that mediates all structures. For Merleau-Ponty, organs such as skin, eyes, and veins live a primordial language (N 222). In his lecture courses on nature, Merleau-Ponty uses the expression "natural" with respect to language to connote meaningful gestures (raising an eyebrow, nodding one's head). Cecilia Sjöholm (2001) notes that the term "natural" language denotes an "even more primordial" sense of language (175). "A sense organ like an eye or a hand is already a language because it fills the function of interrogation and/or response. What Merleau-Ponty calls interrogation, here, is movement and the response is perception, a response to movement" (175). A sense organ's perceptive response could be said to signify to surrounding sense organs the movement it has just perceived. "The body is, in other words, already in a situation where it communicates with itself: touching itself, responding to itself, as well as to other bodies" (175). All but the most simple animals, then, capable of responding to their environments—and not primarily reacting as in the case of the simplest animals[6]—can be said to be a corporeal communication. Such corporeity then, appears to exist *as* communication.

Significant here is not only that corporeity exists *as* communication but that it does *not* exist *of* communication. That is to say, the communication that corporeity exists as is not the basis of corporeity, rather, corporeity is the basis of communication. Without a body with a nervous system, therefore, signification (or communication) appears unable to arise. Thus, communication shows itself rooted in bodily functioning (not vice versa).[7] In Robert Vallier's (2001) terms, "if the manifest activity of the animal body is signification, then all corporeality is already the root of symbolism" (205). We can think of these then—sense organs, gesture, and spoken language—as modes of communication or signification. Whether primordial or complex, any communication "can never be wholly detached from the function of the body" (Sjöholm 2001, 175), and all communication has roots in the body.

PART III. THE INDEFINITE HORIZON OF DESIRE

Not only does all but the most simple animal corporeity exist as and not of communication for Merleau-Ponty, but also as a signifier—a signifier that signifies 'oneself.'[8] Animal behavior if oriented to self-signification, is nevertheless an indeterminate directedness—that is, its intended object remains constitutively indeterminate (PP 55; 111; 135–136). Before showing why animal corporeity is directed toward self-signification, however, it will be helpful to consider in more detail Merleau-Ponty's use of terms such as "oriented to" or "directed toward." This brings us to his concept of desire.

According to Merleau-Ponty, human experience suggests that psychical and physiological functions commingle with one another, almost inextricably along an "intentional arc" (PP 135–136). Terms that are often used to describe Merleau-Ponty's concept of intentional arc—that is, "reaching for," "directed at," "desire for" (PP 55; 11; 135–136)—imply an open and only partially outlined future toward which is aimed. By "desire" for Merleau-Ponty, I mean such indeterminate aiming.

Merleau-Ponty's use of phrases like "oriented to," "directed toward," "desire for," and "projection at" typically are not completed by an object. And yet, each includes a preposition. Traditionally, these prepositions require a prepositional object. That is, a person typically "rises towards some *thing*." Our grammar assumes that an action, like "rising towards," has a determinate referent.

Although the concept of an intentional arc, for Merleau-Ponty, involves a bowing, the bowing is not at a *definite* object or a thing. The logic structuring Merleau-Ponty's concept of intentional arc not only untethers the object from the subject, but diffuses the supposed determinacy of each. If the concept of intentional arc seems to mute or modify a subject-object reflexivity, it nonetheless amplifies an aspect of certain Western notions of self: *eros* or desire.

Merleau-Ponty's concept of desire understood through his idea of intentional arc, in effect, dislodges the traditional framework anchoring most sentences of European descent and the philosophical, religious, and common-sense views of self and goodness built with those sentences. "It is a question of recognizing consciousness itself as a project of the world, meant for a world [or object] which it neither embraces nor possesses, but towards which it is perpetually directed" (PP xvii). Our consciousness never "possesses" objects, suggests Merleau-Ponty. Our consciousness does not experience, according to Merleau-Ponty, objects that can be determined by definite limits. "[S]ensations," he writes, "and images which would be the beginning and end of all knowledge never make their appearance anywhere other than within a horizon of meaning" (PP 15).

We can analyze the latter phrase, "horizon of meaning," by concentrating on the concept of horizon. A horizon can be interpreted as an indefinite boundary. If one considers the horizon of ones' field of vision right now, he or she will probably not see a firm boundary or line demarcating the appearing field from that beyond. Such an indefinite boundary or horizon, Merleau-Ponty suggests, is generally that which accompanies our sensation of any perceptual experience or object (PP 15–18). This would suggest that even if over time the meaning we ascribe to an object appears determinate, the present experience of the object to which it refers is not.

The relation between the English word, "horizon," and the Greek word for definition, ὁρισμός,[9] points the way for a helpful understanding of how definition or limit operates in the context of Merleau-Ponty's work. We can understand ὁρισμός, to mean an indefinite limit, and Merleau-Ponty's concept of intentional arc to stretch toward such a limit. The area toward which it stretches would seem to be an indeterminate "object" of subject-object mechanics. In fact, Merleau-Ponty's concept of intentional arc employs both the language of "horizon" and its etymological heritage implying experience as an indeterminacy of perceptual particulars (PP 22).

PART IV. CORPOREALITY AS DEFINING AND SIGNIFYING SELF AND OTHER

If human orientation directs itself toward an indeterminate object, for Merleau-Ponty, it also shows itself directed toward signifying 'oneself'—a self that shows itself without permanent determination. As such directedness, human behavior illustrates that existing is communication.[10] Vital behavior "is an oriented action, it points itself to something, and this pointing, this indication, is already signification; vital behavior, in virtue of its being-oriented-to, thus *signifies*" (Vallier 2001, 198). In *The Structure of Behavior*, Merleau-Ponty explains that a vital organism's attribute of being-oriented to and signifying is "an attribute of the *perceived* [and perceiving] organism" (qtd. in Vallier 2001, 198). The perceived and perceiving organism includes all but the most simple animals (N 168–173). Merleau-Ponty shows this in his analysis of the zoological studies of J. von Uexküll, E. S. Russell, Adolf Portmann, and others as discussed in notes from his second course on nature (1957–1958).

J. von Uexküll's work is significant for Merleau-Ponty because it shows that animals with nervous systems respond to and orient themselves within their environments. Unlike the "lower" animals, such as worms and sea urchins, they appear opened-up to their *Umwelt* (environment) rather than closed off to it. The lower animals, by contrast, seem not to relate to their environments. Instead, their behavior seems governed according to an anatomical construction (*Bauplan*) that functions much like a machine (N168). So-called higher animals, on the other hand, are characterized precisely by their *relating and orienting themselves* amidst the *Umwelt*. Moreover, higher animals respond to the *Umwelt* as if it were a world of signs (*Merkwelt*). By focusing on the development of the retina in animals, Uexküll shows that "the agency of the exterior world, the objective universe, from now on plays the role of a sign rather than that of a cause" (N 171). The nervous system operates, according to Uexküll, as a means for interrogating, organizing, and responding to forces of the external world (N 171). What is important, for Merleau-Ponty, is that all but the most simple animals are not seen as moving mechanistically but instead, as creatively and in response to signs. This suggests the crucial distinction in Uexküll's concept of *Umwelt*. An *Umwelt* "is not reduced to a sum of exterior events," but is co-created by the conditions of and the animal's regulations to the milieu (N 177). In this perspective behavior of all but the simplest animals is a receiving, from the world, of signs from among which it chooses and with which it shapes to its own tendencies its *Umwelt* (N 177).

For Merleau-Ponty, E. S. Russell's work complements the significance of Uexküll's. Russell discusses the movements of cells in the case of a surface wound. Cells move from deep in the body toward the wound, from the interior toward the exterior—towards the body's beyond. Animal life *differentiates* its interior from its exterior and beyond, with a "directiveness." The implication is a "weak teleology," weak in the sense that it involves a "nonfinality," that is, no determinate end-objective (Vallier 2001, 200–201). Its open-ended finality is a purposive mobilizing from animal interior to animal exterior—even beyond the animal's body. Russell's work is important for Merleau-Ponty because it shows the formation of a condition necessary for the possibility of signification: the animal negotiating a body differentiated from a milieu "beyond" it. This indicates a "sort of presignification" (Merleau-Ponty 1968/1970 & 1988, 163).[11] Reference to an exterior presupposes a sense of "cohesive unity of the animal body" (Vallier 2001, 201). The animal body has to have marked differentiations that can in turn "signal" or "point to" themselves and others *as* themselves and others—whether as itself as an animal whole, or as other external to itself.

If, for Merleau-Ponty, Russell's work helps him to show that animal life *as* differentiation provides a precondition for the emergence of animal life *as* signification, Portmann's work displays such signification's emergence. Portmann stresses the importance of the perceivable *surface* of an animal body. Whereas Russell shows that animal life marks differentiations and shows a directiveness from inside-to-outside, pointing beyond the body, Portmann reveals the significance of the surface of the animal body as an organ for expression and for being recognized (N 187). Portmann attributes to this ability to be seen, what he calls a "value of form" (N 188). Merleau-Ponty interprets this to mean that animals' ability to show themselves as a designated whole to others, especially to others of their species, may have value in itself. That is, it allows perceivers viewing the whole of the animal to behold the "mystery of life" (N 188). "The form of the animal," says Merleau-Ponty, "is not the manifestation of a finality, but rather of an existential value of manifestation, of presentation" (N 188).

Although showing oneself may be in the interest of species survival, Portmann's thesis defends itself against the argument of mere utility. Merleau-Ponty's reading of Portmann points out that were self-display merely an efficient means for species survival, animal mating rituals would probably be less involved than they often are (N 188). "Portmann's study suggests that life is not submitted uniquely to a principle of utility, and that animal appearance and ornamentation (and certain behaviors dependant on these, like the sexual display) have a "value of form" (N 188). For Portmann, the self-showing of animal bodies has a value of form as "an organ for being seen" (Merleau-Ponty 1994, 245).[12] In its appearing, the animal enacts a signifying—it signifies itself, differentiating itself from its surroundings and other creatures.

Merleau-Ponty's analyses of Uexküll, Russell, and Portmann suggest that animal behavior is thus constitutively a process of communication as self-signification. Animal behavior is oriented to creating a sign that signifies the animal *itself*. The sign *for itself* that animal behavior shapes is the animal's own corporeal surface. Uexküll's research provides a concept of animal environment that assumes that an animal perceives a world of signs and regulates them to co-construct its environment with conditions thereof. Russell's research complements Uexküll's by showing animal behaviors that precondition self-signification: the processes of differentiating inner and outer. Portmann's view of the animal-body surface functioning as a sign—a sign of *itself*, to be seen by others, shows the emergence of self-signification. Seen in a context combining the research of Uexküll, Russell, and Portmann, the body's surface can no longer be viewed as merely a static or given sign, but rather one that is hard-won and actively differentiated by the animal itself from its surroundings.

The animal's negotiated surface is that aspect of itself that is perceivable by the perceiving others, which become, at least in part, differentiated from it through the layered process of self-signification described above. Animal behavior in this context can be said to manifest communication not only because it perceives signs that it interrogates and negotiates, but also because it signfies meaningfully; it *directs itself outward* and makes of *itself* a signifier. The signifier is meaningful because it communicates a provisional externality of the animal. The signifier "says," "This is me; this is my space." It furnishes a necessary condition for subsequently pointing to and 'identifying' conditioned others. In this way, bodies can be said to exist both *as* communication and *as* self-signification.

Such existence is accomplished by a perceiver and only exists as such for perceivers. Apart from perception such appearing does not exist. Furthermore, as we can now see, Merleau-Ponty does not view such self-showing/signification as a mere projection of processes that are exclusively human, onto animal life.[13] It is basic to a corporeal structure that constitutes the lived adaptive experience of certain animals and humans. "[B]ehavior can be defined only by perceptual relation and that Being cannot be defined outside of perceived being" (N 189). Although Merleau-Ponty hierarchizes orders of animals in *The Structure of Behavior* (1983), by the time of his lecture courses on nature, he suggests a horizontal, co-communicating "empathy" between human bodies and animal bodies (Vallier 2001, 205).

Merleau-Ponty illustrates that self-signification is basic to a corporeal structure constituting the lived adaptive experience of all but the simplest animals.[14] What is significant about this notion that all but the simplest animals live oriented towards a self and other designation is that it shows that corporeity is the root of communication and demonstrates that existing is, itself, communication.

PART V. TRAUMATIZED BODIES

Thus far I have emphasized Merleau-Ponty's analyses of animals. These focus on nonhuman animals but inferentially extend to humans. In the *Phenomenology of Perception* (1945/1962), Merleau-Ponty's analyses emphasize, by contrast, humans. There, Merleau-Ponty appeals to cases of brain injury, paralysis, and limb amputation, among others. Merleau-Ponty draws upon neurologist Kurt Goldstein's[15] empirical examinations of the radically shifted worlds of brain-damaged combat survivors from World War I. Such worlds originate and persist, Merleau-Ponty writes, because of "one present among many"—the happening of a traumatic event—"which thus acquires an exceptional value" (PP 83). By examining the physiological and psychic phenomena that arise in the aftermath of traumatic experience, Merleau-Ponty magnifies human corporeality as a directed self-showing. For Merleau-Ponty, trauma operates like a prism through which the body as it is experienced, lights up.

Given this special position of the phenomenon of trauma in *Phenomenology of Perception*, it surprises me that not more has been said about it. In interpretations of Merleau-Ponty's work, attention to the category of trauma is virtually absent, even if tangential categories like repression, violence, and unconsciousness are recurrent themes.

For Merleau-Ponty, the brain-injured and amputees of Kurt Goldstein's studies, if traditionally understood as survivors of so-called physical trauma, are also survivors of psychological trauma. Although Merleau-Ponty indicates the exceptional status of trauma for his project, he does not explicitly explore, specify, or reveal his implied concept of trauma. Here I will begin such an analysis. I will explore three kinds of bodily adaptation: anatomical, technological, and verbal—as expressions of preconscious bodily desire and especially of that which is arguably a human being's most intense kind of preconscious desire—that generated by traumatic experience. By focusing on trauma—bodily desire at its most extreme—I show the remarkable extent to which Merleau-Ponty is able to make corporeality visible as directed toward self-signification that both expresses and sinks its roots in corporeity.

PART VI. ANATOMICAL ADAPTATION

Persons injured during combat often experience and respond to trauma across several interpenetrating levels of embodied adaptation.[16] I will provisionally divide these into three and call them anatomical, technological, and verbal adaptation. Merleau-Ponty's work generally offers little developed analysis of what I call "anatomical adaptation."[17] By "anatomical adaptation" I mean the spontaneous mobilizing of an organism's body cells, tissue, and organs, in response to preconscious felt needs. If anatomical adaptation is rarely scrutinized in the *Phenomenology of Perception*, it is nonetheless alluded to repeatedly via Merleau-Ponty's many references to amputated limbs, injured brains, and paralyzed bodies. The analysis of anatomical adaptation that I offer below is my own. I limit the discussion to one example of a traumatized body to which Merleau-Ponty refers—the paralyzed body. In particular, I will focus on recent research on the spinal-cord-injured body.

In cases of paralysis from spinal cord injury, the anatomical body marshals a complex set of internal adjustments to strike an equilibrium with its new circumstances. Among persons with cervical spinal cord injuries, that is, above the spinal segment range of thoracic one to lumbar two or three (T1 to L2 or L3), a process of equilibrating unfolds over a period of weeks, months, and years. The spinal cord injury alters the functioning of many corporeal processes and one such process is the autonomic nervous system—the part of the vertebrate nervous system that governs automatic internal processes (Somers 2001, 32–33). Spontaneous processes like cardiovascular function, internal-temperature regulation, secretion (sweat, saliva, urine), and intestinal contraction (enabling excretion) begin operating under the extremely altered circumstances caused by the injury (Bullock 1996, 451–455). Within the first several weeks after the injury, persons with injuries above the T1-L2/L3 spinal range (an area that governs the sympathetic nerves of the automonic nervous system), are prone to experience bradycardia (slowing of the heart), bradyarrhythmia (alteration of heart beat) and hypothermia (Somers 2001, 32–33).

For our purposes I will focus on the condition of hypothermia. In atmospheric temperatures that would typically feel comfortable to one, the recently spinal-cord-injured person is likely to be chilled and is vulnerable to hypothermia (Somers 2001, 32–33). This is because the person's sympathetic nerves have stopped functioning. Such loss results in the peripheral veins dilating (according to directives from the parasympathetic nervous system located at the brain stem and so, above the level of any possible spinal cord injury) without the corresponding reflex of constriction normally governed by the sympathetic nervous system (Somers 2001, 32–33). Dilation causes loss of heat to the spinal-cord-injured body especially for several weeks after the injury (Somers 2001, 32–33).

Significant in my view is the adaptive process of the sympathetic and parasympathetic nervous systems within weeks of the injury. After about three weeks, spinal-cord-injured persons stop experiencing an extreme tendency toward hypothermia. This is because of a dual adaptive process. The sympathetic spinal nerves begin to function again—although at a reduced level—despite broken communication between them and the parasympathetic nerves. Also, at about this time, the parasympathetic nerves begin to reduce their output, causing a disproportionate dilating effect to compensate for the compromised effect of constriction given the compromised sympathetic nerves. A gradual process of anatomical shifting sets in. After the injury, one's spinal nerve anatomy shows itself vigorously exploring and attempting to survive the new circumstances. Through a process of interrogation and response, the autonomic nerve processes of the spinal-cord-injured body gradually reorient themselves and adjust. Such adaptation could be said to exhibit what Sjöholm calls "primordial 'natural' language" (2001, 175).

By moving (interrogating) and perceiving (responding) the compromised autonomic nervous system *creates* a new strategy for temperature regulation, given the altered conditions. If it succeeds, it will have equilibrated itself relative to its new environment. Although persons with spinal cord injuries remain more susceptible to both hypothermia and hyperthermia (due to loss of sweat glands[18]) even after the above-described autonomic nervous system adaptations, they are much more comfortable and thermodynamically equipped to survive than they were during the first three weeks following the injury. Significant for our argument is that the autonomic nervous system shows the capacity of mammalian anatomy to negotiate its place and reconfigure itself according to the new circumstantial demands.

If the traumatized autonomic nervous system makes its most steep show of adaptation during the first few weeks after the injury, its adapting and that of other anatomical processes continue in noticeable ways for roughly the first two years after the injury. Generally during this time persons perceive a gradual increase of overall bodily ease of functioning and comfort. Several years after his accident, Christopher Reeve's body, though still more susceptible to heat and cold than a non–spinal-cord injured body, and barring other complications, had become more efficient and felt for Reeve more comfortable than during the first six months after the injury.

Skin adaptation is another example of anatomical shifting and a consequence of trauma to the spinal cord (Somers 2001, 32–33). Unremitting pressure on muscle and skin tissue cuts off blood circulation to the tissue and the tissue dies. People with healthy spinal cords spontaneously shift their sleeping and sitting positions countless times during a day, in part because their bodies are responding to felt needs of which they are generally unaware. While sitting they relieve pressure in one area by shifting it to another and so on. Quadriplegics spend hours sitting on their buttocks in wheelchairs or lying on their backs while sleeping. The spontaneous posture shifts are not possible. Christopher Reeve spoke of skin breakdown in his body so extensive that one could fit a fist inside one wound where skin and tissue had died (Rosenblatt 1996).

Here, as in the function of temperature control, the threat is strongest during the first months after the accident. Although skin resilience and adaptation is unique for each body, it is common knowledge among the spinal-cord injured that, after several years, the live tissue of some paralyzed bodies adapts to sitting or laying for many more hours in one position in a wheelchair or bed than it initially could do. In some cases, bodies of persons with spinal cord injury have transformed part of their live tissue into thick dead callus, in combination with certain other internal tissue adjustments, to defend against the internal spread of dead tissue.

Although the self-demarcating, reorienting aptitudes of more simple anatomical mechanisms, such as autonomic nerves and skin tissue, may be less expansive than those of more complex mechanisms (cerebral neurons), they are also more expansive than has been traditionally believed. As James Grau and Robin Joynes (2001) show, in recent studies about Pavlovian and instrumental conditioning within the spinal cord, the spinal cord neurons can learn and remember, albeit in simple and more limited ways than cerebral neurons. Grau and Joynes's conclusions are based on experiments with rats whose spinal cords have been transected. Various experiments pairing environmental cues with an unpleasant stimulation (electric shock) show that spinal neurons can learn the relationship between a certain environmental circumstance and pain/danger (letting one's leg rest in a salt solution and the corresponding electric shock when it does so). The spinal neurons can "learn" and execute a leg flexion movement that lifts the leg out of the solution, so as to avoid the shock. Grau and Joynes's "aim has been to characterize the functional mechanisms that underlie learning within the spinal cord" (2001, 46). Their work "confirms that these systems are sensitive to the environmental relations that underlie Pavlovian and instrumental learning" (46). This sensitivity however lacks the more complex learning capacity of cerebral neurons that can "map new environmental relations across time irrespective of the stimulus modality, response form, or affective state" (46).

Even if the spinal cord behaves and learns more simply and with more constraints than the brain does, Grau and Joynes note that its aptitudes have nevertheless been underestimated. According to traditional views, learning requires consciousness and/or the more complex networks and neurons of the brain. In other words, learning has been believed to happen above the spinal cord, not in it or in other noncerebral anatomy. Recent studies by Michael Patterson (about spinal fixation in animals),[19] Jonathan Wolpaw (about spinal stretch reflex),[20] and Anton Wernig, Andras Nanassy and Sabina Müller (about treadmill therapy for para- and quadriplegics)[21] further support the notion that spinal-cord neurons learn and remember. Such developments in spinal-cord research are significant for Merleau-Ponty studies. They corroborate Merleau-Ponty's view of animal behaviors as interpenetrating co-responses to felt bodily needs, not self-contained faculties with defined functions.

Examples of spinal-neuron, autonomic-nerve, and tissue adaptation in spinal-cord–injured bodies are significant because they show anatomical processes behaving as communication—interrogating and responding to felt needs. They are reminiscent of Sjöholm's view (2001, 175) that a form of primordial language is implied by Merleau-Ponty when the latter analyzes the movement (interrogation) and perception (response) of cellular anatomy (2003, 219). The extreme cases of spinal cord trauma evince anatomical processes *as* an organism's orientation toward self-signifying. The gradual process of adapting to the altered conditions for controlling body temperature and maintaining live tissue shows anatomy enacting and participating in the structure of lived behavior as an "intentional arc." They support the idea that animal behavior for Merleau-Ponty exists *as* communication, with a weak teleology aimed toward self-designation or the making of oneself a 'spectacle' to be seen. One sees this desire, to negotiate with one's surroundings a territory of 'self,' in the gradual processes undergone by the spinal-cord injured. The process of anatomical equilibration of spinal-cord–injured bodies magnifies anatomy as a behavior interpenetrating with other behaviors. Anatomical adaptation in the case of extreme trauma reorients and recreates anatomical relations in the face of new circumstances. The gradually emerging, newly equilibrated anatomy of the spinal-cord–injured person underlines the idea of anatomical motility as a self-showing, differentiating self from other.[22]

The existence of this designating of self, shows itself rooted in interrogation of and response to preconscious felt need or desire. The anatomical shifting of the spinal-cord injured, shows anatomy redesignating the contours of anatomical 'self' in the sense of newly adapted mechanisms for temperature control, and new rules apportioning callus tissue and live tissue. The equilibrating anatomy expresses the body's desire to continue life as a self-showing that self and other signifies. It simultaneously differentiates itself from external others and points to itself and others. It signifies itself and others as selves and others, both to itself and to perceiving others. The example of anatomical shifting is important because it shows signification or communication operating physiologically below and independent of the cerebral cortex. Moreover, the process of anatomical shifting vigorously negotiates the boundaries between self and other, life and death that make possible one's ability to signify and display one's lived body as a symbol for oneself and others. The most significant consequence of Merleau-Ponty's view of corporeity existing as communication, however, is that communication thus reveals itself to be of and dependent upon bodies and bodies of and imbedded in nature. This view effectively reconnects that which much philosophy since Plato and the rise of literacy in the West disconnects: idea and matter, spirit and nature, animals "with language" and animals "without language."

If anatomical adaptation both shows life and participates in life as directed toward such self and other signification, the same might be said of technological adaptation. Walking sticks, contact lenses, hearing-aids, wheelchairs, and other assistive technologies show themselves likewise of and expressing bodily needs. In co-participation with anatomical adaptive processes, technological behavior, as I show below, also manifests oneself as a desire to self-show—or, again, as communication.

PART VII. TECHNOLOGICAL ADAPTATION

If in countless experiences of anatomical need, body organs, and cells interrogate, signify, and adapt—in countless others, their aptitudes for response are too limited. When felt needs of humans remain unmet, human corporeity often shows itself expressing and responding via assistive technologies. By "assistive technology" I mean technologies built specifically for a person with a certain disability in order to fulfill a certain task that would typically be manageable independently by persons without the disability. These technologies extend our *lived* bodies beyond our anatomy. The example of paralysis underlines the status of such built instruments as rooted in and expressing bodily desire.

Two key technologies for the quadriplegic are manual and electric wheelchairs and voice-activated home environment systems (i.e., thermostats, lights, and computers). The felt need to be self-mobilizing and in control of one's home environment has given rise first to *imagined* and subsequently *built* rejoinders to such needs. Imagined technologies (the idea of an electric wheelchair) and actualized technologies (an actual electric wheelchair) express and help meet felt needs.[23]

Living with such technologies extends the lived body of the spinal-cord-injured person. One's habits, health, and daily options are opened up proportionately by the powers afforded by the technology. Having grown accustomed to his or her wheelchair, the person's dimensions reflect those of the chair. The chair is not an object between oneself and other objects, but just as one's given body extends towards self-showing, the chair facilitates such extension. The relation between one's paralyzed body and chair is analogous to that between Merleau-Ponty's blind man and his stick. "The pressures on the hand and the stick are no longer given; the stick is no longer an object perceived by the blind man, but an instrument *with* which he perceives. It is a bodily auxiliary, an extension of the bodily synthesis" (PP 152). The stick and chair are not objects for the blind man or paralyzed person respectively, but instruments that extend their perceptual experience according to the aptitude of the technology. "The blind man's stick has ceased to be an object for him, and is no longer perceived for itself; its point has become an area of sensitivity, extending the scope and active radius of touch" (Merleau-Ponty 1962, 143).

Some persons in wheelchairs speak of being able to discern with immediate precision whether their body and chair can make a particular turn, fit through certain doorway, and so forth. His or her lived body-and-chair requires a certain area width and length when traveling straight ahead, and a certain radial-space when turning a corner. A person in a wheelchair "feels" where the chair begins and ends "without any calculation," just as a person without a chair goes "through a doorway without checking the width of the doorway against that of [his/her] body" (PP 143). Accordingly, rooms and options become perceivably open (or closed) to the lived body-cum-chair of the spinal-cord injured. The wheelchair's dimensions become in a certain sense one's own.

Assistive technology points to a self-showing that is also an expression of bodily desire. In the case of spinal-cord–injured persons, the assistive technology reconfigures the trajectory of one's signifying; it allows a person to continue *life as expression* of given and acquired habits, habits that are or are becoming 'self' and 'other' differentiating and signifying. The assistive technology both signifies human bodily desire and participates in a spinal-cord–injured person's reorienting and redefining the boundaries of self and other. Together with an adapting anatomy, assistive technology expresses one's lived body as oriented toward a self-showing, the purpose of which is to create, mark, and sustain designation and differentiation of self and other.

The lived body of a person prior to such an injury marks, enacts, and signifies itself without such assistive technologies. During the initial years after the injury, he or she reorients, recreates, and redesignates his or her corporeal self and *surface*. One's corporeal self as a 'spectacle' to be seen by others—what Portmann calls a "value of form"—has radically 're-presented' itself. The assistive technology (for instance, the electric wheelchair) becomes a part of the spinal-cord–injured person's lived corporeity. It exhibits the spinal-cord–injured person and his/her community, mobilizing "the given world in accordance with projects of the present moment to build into a geographical setting a behavioural one, a system of meanings outwardly expressive of the subject's internal activity" (PP 112). Like adaptive anatomy, adaptive technology allows the spinal-cord–injured body to continue enacting itself—if differently than before the injury—as a self-showing that actively draws, decides, and points to itself and others. If the example of anatomical shifting returns communication (and thus, too, language) to physiology, the example of technological adaptation extends bodies beyond physiology and shows how one's self-designating self-showing adapts to include the technologically appended lived body.

PART VIII. VERBAL ADAPTATION

Speaking

Anatomical and technological adaptation show themselves as two of many processes through which human corporeality signifies both itself and a beyond-self; for Merleau-Ponty, a third such process is speech. By "speech" I mean spoken or written language that accomplishes through words, thought.[24] The *Phenomenology of Perception* makes explicit the body as the basis of speech.

Speech, like anatomy and technology, shows itself as signification *of* and intention toward *expressing* bodily desire to demarcate and signify self and other.

Speech, for Merleau-Ponty, is not only inseparable from corporeity but emerging from and existing *as* corporeity, especially *of* and *as* desire. To see this warrants our looking at the relationship between speech and thoughts, and the meanings associated with them. From the *Phenomenology of Perception*, we are familiar with Merleau-Ponty's idea that the meaning of speech is not separable from words. That is, unless one revisits and repeats a thought already conceived, one typically does not know where one's words are going while one is saying them. For Merleau-Ponty, words do not "clothe" thoughts that are separable and conceivable independent of the words (PP 182). Thought happens through and as words, and meaning is accomplished through and as speech (PP 178).

Just as the "musical meaning of sonata" does not exist apart from the notes, thought or meaning of words does not exist apart from the words (PP 182). "Thought is no 'internal' thing, and does not exist independently of the world and of words" (PP 183). Not only is thought inseparable from speech but both are inseparable from corporeity. Merleau-Ponty suggests this when he singles out a provisional concept of "'pure' thought." He conceives speech or thought in their "pure" form *as* desire and "void of consciousness" (PP 183). "Just as the sense-giving intention which has set in motion the other person's speech is not an explicit thought, but a certain lack which is asking to be made good, so my taking up of this intention is not a process of thinking on my part but a synchronizing change of my own existence, a transformation of my being" (PP 183–184). Significant then, is that speech or thought shows itself to be born of a certain prereflective "lack" or wanting. *It emerges of desire to create sense in conditions where none yet is. It is born of a situation in which there is initially only "intention" (desire).*

Speech and thought show themselves to be not only inseparable from, but generates of and as bodily desire. Importantly, "'pure' thought," for Merleau-Ponty, does not involve attributes traditionally associated with thought: representation and consciousness. "'Pure' thought" is said to be qualitatively desiderative and unconscious. Thought or speech in its "purity" begins as an intention or desire—to give sense.

Speaking the Unspeakable

Now, for Merleau-Ponty, traumatic experience "acquires exceptional value" because the preconscious desire or "pure thought" constituting the "trauma present" is enormous. It overwhelms and demands attention for itself long after the actual event has occurred. Merleau-Ponty argues that traumatic memory images, like all other memory images and behavioral processes, participate in structuring the direction of one's movements (interrogations) and perceptions (responses). Unlike "normal" memory images and processes, however, a traumatic memory-image's tendency toward expression is immense (PP 83). Other ordinary present experiences, by comparison, command comparatively little intention toward expression.

One reason why humans find it experientially difficult to integrate the supposed memory of catastrophic experience with that of ordinary events is underlined by the etymology of the word *trauma*. Its German cognate, *Traum* (dream), points to a long-standing disjunct in human experience between integrated memory and disintegrated memory. And, like dreams, the experiences one undergoes during a traumatizing event generally remain segmented off from conscious memory. Clinical psychologists often call such disintegrated memory images, "dissociative memory."[25]

Survivors of life-threatening events, or trauma, tend to experience the event and re-member it afterwards through images that are generally dissociated from ordinary memory. A traumatic experience—often life-threatening—is, says Merleau-Ponty, so unusual that the images one experiences during the event, and associates with it afterwards, are placed outside the bounds of normal memory (PP 83). They hide out, one might say, absorbed in a condition of repression. But although these images abscond, they exert an ongoing pressure on our ordinary memory-images and daily experience. The traumatic experience

> . . . acquires an exceptional value; it displaces the others and deprives them of their value as authentic presents. We continue to be the person who once entered on this adolescent affair, or the one who once lived in this parental universe. New perceptions, new emotions even, replace the old ones, but this process of renewal touches only the content of our experience and not its structure. Impersonal time continues its course, but personal time is arrested. Of course this fixation does not merge into [ordinary] memory; it even excludes memory. . . . (PP 83)

The hiding, hard-pressing, displaced trauma-images demand attention while remaining steadfastly beyond view. One's trauma memories remain "constantly hidden behind our gaze instead of being displayed before it" (PP 83).

Merleau-Ponty discusses traumatic memory, but he does not develop an analysis of trauma with respect to memory or speech. To show that the so-called "representation of traumatic experience" is what I call "adaptive speech"—incidental to and a betrayal of traumatic experience, I will offer a revision of psychiatrist Judith Herman's Chapter, "Rememberance and Mourning" in *Trauma and Recovery: The Aftermath of Violence—From Domestic Abuse to Political Terror* (Herman 1997).

As Herman sees it, "The fundamental premise of the psychotherapeutic work is a belief in the restorative power of *truth-telling* [my emphasis]" (1997, 181). I should like to argue that such "*truth-telling*" appears to be experientially impossible, but that the activity of *creating* an idiom for a *new* narrative tale inspired by the wound is not. The created tale, insisted upon by the relentless demand of the preconscious traumatic wound, fails however to represent the wound or trauma experience itself. The traumatic wound, I argue, excludes linguistic representation[26] especially in narrative form.

Herman's own essay implies this impossibility. The survivor is encouraged to tell the true story of her/his terror via an arduous and demanding "flooding" technique. This includes a reliving of the affects and thoughts experienced during the traumatic event, and an examination of "the moral questions of guilt and responsibility" in order to ascribe not only context, fact and emotion to one's tale, but also meaning (Herman 1997, 182). The flooding process is called by Herman "controlled re-living of the experience," but in fact, it exposes patient and therapist to more (or, perhaps more accurately, to less) than the unsublimated trauma experience itself. Not only are the "fragmented components of frozen imagery and sensation—the smells, sounds, a pulsing heart, constricting muscles, alternating emotions—of revulsion, confusion and fear—included in this tale, but also included is a verbal retrospective, with an order and meaning attributed in hindsight (Herman 1997, 177). For Herman, truth- and storytelling engages the survivor emotionally, socially, and religio-philosophically. However, in the case of survivors who are fortunate enough to be at this second stage of Herman's three-stage recovery scheme, their emotional, social, and metaphysical orientations are admittedly in the throes of a co-creative self-reconstruction authored by the survivor and nurtured by the therapist. Visible and significant, if unrecognized by Herman, is that the retrospective narrative of the trauma involves modes of human experience (narrative, representation, organization, meaning coherence) that were and are generally absent from the traumatic experience itself. Put differently, complex communication (language) and neural processes are generally stunned and overwhelmed during a traumatic event (van der Kolk 1995, 172–176). They are generally not able to interact with nor integrate the event as usual—in part because this is no mere usual event—and this may explain why traumatic memories are often dissociated from ordinary memory (van der Kolk 1995, 172–176). Language is not able to be present for the event. In this perspective, the trauma as experienced virtually excludes language. The "true story" that Herman says demands telling and can, she says, be told, appears more precisely to be a paced eruption—into the form of speech—of corporeal desire driven by the power of the traumatic wound, but not representing it, or approaching a telling of "the" story, let alone "truly" so.

The forms of speech into which the wound's demand ex-presses itself are of the wound in only one sense—however important it may be. As Merleau-Ponty compellingly illustrates, all speech begins as preconscious desire (in this case the desire of or as the wound). For Merleau-Ponty, directedness (or desire) as such shows itself as the origin of language and speech, even if it nevertheless remains absent of language and thus too, of speech. That is, bodily desire *as* presupposed drive for spoken expression (that has not yet become expressed) is at its root constitutively unreflected, nonverbal, and unmeaningful (PP 183). Speech, then, emerges of and as the bodily desire to create sense amidst conditions wherein none yet is. In the case of recollecting trauma, one has impaired access to linguistic memories of the event because language was impaired during the event. This impairment together with the prereflected origin of speech is significant. It shows that the retrospective trauma narrative—as something reflected, represented and meaning-ascribed—is *incidental to the traumatic experience itself.* Gregg Horowitz corroborates this view in his description of uncontrolled, intrusive trauma memories (as opposed to therapeutically nurtured ones). "[W]hen the traumatic event recurs, it does so as itself, that is, as unsublimated; it thus confronts the sufferer as what cannot be represented as other than it is, which is to say, as what cannot be represented. . . . At the heart of present forms of the past's future, the past juts out; in remaining unmediated by the available forms of mediation, traumatic insistence is the ruination of the representation relation" (Horowitz 2001, 124).

I should like to suggest that *speech, understood to be an expression of its origin—provisionally "pure desire"—shows itself to be less a mode of representation than of bodily adaptation.* What I mean by "speech *as* bodily adaptation" can be illumined by the two parallel examples of bodily adaptation: anatomical adaptation and technological adaptation. We can recall Russell's discussion of cell movements in the direction of a surface wound. The cells show the autonomic processes of anatomy moving (interrogating) and perceiving (responding to) internal and external conditions—promoting survival and self-signification. The directedness of these cells simultaneously show bodily desire constructing/negotiating its bodily boundaries and so, creatively differentiating and signifying itself as a self-designated (dynamic) whole with respect to its environs. Anatomical adaptation shows animal life *as* directed toward such self and other signification not merely for the sake of survival, but also for the sheer making of oneself a spectacle to be seen. Use of contact lenses, prosthetics, and other assistive technologies do likewise. Both anatomical and technological adaptations show the lived body actively responding to preconscious bodily impulses, and in extreme cases, aggressively negotiating bodily survival: for instance, redrawing (in cases of traumatic bodily injury) and differentiating its lived corporeal boundaries (including prosthetics, the blind man's stick, and the wheelchair).

Adaptive speech, like adaptive anatomy and technology, shows itself of and expressing bodily desire and likewise the preconscious need to stabilize, demarcate and signify oneself. It is an example of the lived body actively creating and negotiating the boundaries between self, others, and surroundings. Herman's work implies—but does not recognize—that speech in fact registers *adaptation*—not representation—with respect to traumatic experience. Although Herman does acknowledge that horrific events destroy the self or the whole, she does not recognize that with the destruction of the preterrified self goes also any proper representation of the horror in its aftermath. Although she recognizes that the ability to cope with the traumatic experience requires developing "a new self" (Herman 1997, 196), she does not acknowledge the incongruence between the retelling of the tale by the new self and the 'experience' of the trauma itself, closed off from the complex communication processes appealed to in its aftermath.

Of specific importance in the retelling by a new self is that the "truth telling" that Herman speaks of substitutes for what can be more appropriately understood as adaptive speech, which itself appears rooted in prereflective bodily desire. In the aftermath of trauma, it seems that speech—like the anatomical and technological bodily processes discussed—adapts to the demands of traumatic experience by renegotiating the conditions of one's self-understanding and self-signification. This radical redrawing of self requires not only that the survivor reconnect with the new self-delineated boundaries once they are accomplished, but that she reconnect with others as a newly drawn self. Importantly, if one's recreated self makes use of verbal and reflective resources, it does so not because the new self is rooted in them. Like anatomical and technological adaptation—which likewise manifest bodily desire *as* communication and self-showing as their manifest activity—the signification of adaptive speech, or in this case of traumatogenic adaptive speech, is rooted not in the communication it shows but in corporeal desire or the intense demand of a prereflective wound.

Disclosing the source of "truth-telling" as the preconscious call of a suffering body does not render Herman's work unimportant. The disclosure is significant because it sustains Herman's ideas about trauma narrative by *reorienting* them through the recognition of Herman's own blind spot. By misidentifying traumatogenic-adaptive speech as a true "telling of the trauma story," Herman dismisses and remains (at least technically) in denial of the enormity and opacity of traumatic experience. She veers toward a certain "triumphalism" against which Gregg Horowitz has warned us (2001, 130). Such triumphalism bears symptoms of denial of—or a wish to complete—one's mourning; it displays, suggests Horowitz, a symptomatic wish for the dead to stay quiet (2001, 122).

As Horowitz's words would suggest, Herman's optimism and overestimation of the power of representational speech underestimates horrific experience and relegates it to a potential but false mastery. Indeed, Herman's clinical experience if not her methods for interpreting it, shows the distance of dissociative trauma images and affects from representational form. Her interpretive choices nevertheless assume traumatogenic experience to be conformable to ordinary perceptual and reflective bounds, even when it shows itself exceeding them. The trauma as trauma, bursts out of any human habitudes and perceptual structures of reflection and linguistic representation. And so, even if a newly created narrative corresponding to the event can be told—and Herman's work suggests as much—*the newly created or* ***new*** *trauma story remains other to the wound or traumatic experience itself.* Extreme terror appears as other than human representation.

If we want to speak of keeping "true" to the trauma (and perhaps we do not?[27]), or of speaking in such a way that more probably attempts "health"—if falling short of what Horowitz calls "the ability to live mournfully," *then perhaps we ought say that traumatic experience is "told" truly as unspoken and unspeakable*: and this is not to say that we ought to limit ourselves to silence, but that we should not mistake the adaptive stories said in the aftermath of our dead, for more than contemporary modes of response to prereflective calls of wounds fated to remain unrepresentable. In the film by Marguerite Duras and Alain Resnais, *Hiroshima mon amour*, the character of the French actress, in dealing with the legacy of her tragically ended wartime love affair, avoids such a mistake about contemporary responses to an overwhelming wound. This nonetheless does nothing to ease her torment. She eventually gives in to her desire to tell that which she knows cannot properly be told; her sacred horror becomes desacralized, and she writhes in recognition of the horror's excess.[28]

PART IX. INVISIBILITY OF BODY

If the traumatized body magnifies behavior *as* being-oriented-to self-signification rooted in and expressing bodily desire, it also reveals traumatic memory, itself, to be excessive to ordinary modes of reflection. Traumatic memory as traumatic is not made visible, but remains opaque. This sort of invisibility—that is, of traumatic 'experience' per se—dynamically engages with the movement toward visibility and self-signification of the body. Invisibility of the lived body, which is brought about through extremity, ought not be confused with another sort of corporeal in-transparency, brought about through the ordinary. I am referring to the perceiving body as medium and measure of perception, which, for Merleau-Ponty, necessarily exceeds any self-showing of itself to itself. Certainly, were there a possibility of perceiving lived body in full transparency, such a perception would have to make such experience visible to oneself not only as a perception but also as the set of conditions organizing such a perception. A full transparency of the lived body, for Merleau-Ponty, contradicts, of course, conditions enlivening the body.

Such excess to visibility—or opacity—is worthy of our attention so that we might properly distinguish it from the excess and opacity of trauma. As a perceiver intending toward self-signification, the lived body can be said to make or "invent"[29] the visible. The making of the visible presupposes a condition in which the visible has not yet begun to be made, and therefore too, presupposes a condition of the visible beginning-to-be-made. If the live body "is the power to invent the visible" (N 190), then it is also a process constituted at its basis by starts. Every perception emerges and differentiates itself from such a state of nonexistence, that is, of not-begun-to-be-made. And "[p]hilosophy," too says Merleau Ponty, "is not the reflection of pre-existing truth, but, like art, the act of bringing truth into being" (PP xx). Given that an aspect of what I call "beginning" participates in the formation of each perception, philosophy of the lived body, to be "faithful to its intention," must carry on toward a seeing that has always already not yet begun; it must, as Husserl writes, carry on "never knowing where it is going" (qtd. in PP xxi). Or, put differently, never properly representing where it has been.

Thus, a condition necessary for the possibility of the beginning of an instance of visibility, for Merleau-Ponty, is invisibility. An emergence of visibility presupposes the invisibility, out of which it rises. Visibility itself, is of and a differentiation from invisibility. The body that 'shows' itself to us as such a 'not-showing' reveals itself via deduction. It indicates its invisibility to be the very structure that conditions the appearing of appearances. The body as invisible ought not be thought of primarily as "absence"—absence of visibility—even though it is such an absence too—but as the corporeal structure that conditions visibility. That is, it is the structure that makes perception happen.

Significantly, although the traumatized body magnifies the body as a process of making visible that self-signifies, it does not obviate the corollary structure of live body that remains invisible (imperceptible). The live body as invisible remains unseen for two reasons: first, because as the conditioning structure for perception, it is itself in its provisional 'purity' absent of any perceptual particulars. And, second, because it is the structure conditioning the rise of actual perceptions, it cannot be seen amidst such conditioned particulars as it is in itself. As the source of perceptual particularity, it is not itself a perceptual particular. To be seen, it would have to be mediated through perception as a perception. In this sense, the body with respect to its conditions for perceptions remains invisible. "It preconditions our involvement with the world, but as such, it is not available to any kind of knowledge that is not filtered through that involvement in turn" (Sjöholm 2001, 180). One's body is not just any structure, but the structure conditioning all perception and experienced structures. It thus holds a special place with respect to visibility; corporeality appears both visible and invisible, but in different respects. It shows itself in one respect as perceived/visible, and this perceived body shows itself demarcating and signifying itself and other(s). In another respect, it 'appears' as invisible. This invisible body structuring the conditions of all perceived structures cannot make itself visible except as mediated—which is to say, except as that which it is not. In this respect, the body remains opaque. If certain traumatic wounds sustain their intractable fate due to the extraordinary horror that bore them, the lived body sustains its originary invisibility due to the necessary and ordinary structures that condition all perceived structures. Thus, the lived body is, in some respects, compelled to remain opaque of ordinary necessity, and this obscurity is of an altogether different ilk than that of a traumatic wound. The latter, by contrast, razed from the extraordinary necessity of trauma, unleashes in one the call of a present suffering—unanswerable and ongoing.

PART X. LIVING WITH PERSPECTIVE ENTAILS THREE ASPECTS OF EMBODIMENT: SITUATIONAL, MNEMONIC, AND INVISIBLE

This lived body 'showing' itself to itself as both visible and invisible in different respects, also implies an accompanying behavior that I call "perspective." Understanding this perspectival behavior will help us compare Merleau-Ponty's thought with Nietzsche's. Lived bodies, for Merleau-Ponty, live with a perspective. The living with a perspective presupposes living as or from a certain bodily orientation. Embodiment not only coincides with a certain time historically but also with a certain place spatially. This does not mean, for Merleau-Ponty, that having a perspective reduces to a spatio-temporal materialism. It means that living involves several aspects of embodiment. For the sake of simplification, I will divide these into three.

The first aspect involves living a body as a "particular situation" (Sjöholm 2001, 173). Cecilia Sjöholm uses this phrase to describe the fantasy illustrated in the film *Being John Malkovich* (2000). A person accustomed to living with a particular musculature, facial appearance, height, foot size, and so on, wakes with the corporeal physical details of another person. That one lives a particular height, hair color, muscle arrangement, and bone size expresses an aspect of embodiment that can be called "a situation of embodiment" (Sjöholm 2001, 173). The fantasy of waking up with another's embodied situation provisionally distinguishes an aspect of embodiment that is a necessary condition for the possibility of living with a perspective. It can be differentiated from a second aspect of embodiment that presupposes situational embodiment.

This second aspect I will call "mnemonic embodiment." Mnemonic embodiment involves types of memory: cellular memory, muscle memory, representational memory. And, it is also a condition for living with a perspective.[30] If we continue with the example of *Being John Malkovich* (2000), we can imagine an alternate fantasy, this time illustrating mnemonic embodiment. Gymnast Mary Lou Retton awakens in the body of John Malkovich. She would be able to run through her balance-beam routine in her mind, imagining the timing and sensations of actually executing the routine as she has hundreds of times before. Encumbered however, with John Malkovich's different muscle tone, bone size, and physical arrangement however, she would be at serious risk were she to attempt the routine. Having a perspective presupposes fundamentally living a body, and living a body can be described in terms of three aspects. The first two of these aspects of living a body, situational embodiment and mnemonic embodiment, have the quality of visibility (or perceptibility).

The third aspect of having perspective involves living a body as invisibility, and is an aspect of embodiment that is presupposed by both situational and mnemonic embodiment. This sense of embodiment I described earlier; it is the ordinary invisibility of the body. Such "invisible embodiment" is not the experience of visible particulars of one's body, but it does not happen without such particulars. It is instead an experience of invisibility. This is the opacity of the structure of one's body that conditions all experienced structures (i.e., perceived particulars). Experience of this structure cannot become visible to one except as a not-showing-itself-as-itself. Experiencing one's body as the structure that conditions one's experience of all perceptions is, therefore, an experience of not-seeing. If living a body, for Merleau-Ponty, is to experience oneself in different respects as both visible and invisible, it is also to live with a perspective. Living with a perspective can be said in turn, to presuppose three aspects of embodiment: situational, mnemonic, and invisible.

PART XI. CONCLUSION

Adaptive Anatomy, Technology and Speech

This chapter analyzes recent developments in the areas of spinal-cord injury and trauma studies to underline Merleau-Ponty's view of animal behaviors (i.e., communication and desire) as interpenetrating one another and the milieu. By bringing into dialogue new ideas in spinal-cord, trauma, and Merleau-Ponty studies, the chapter offers a new framework for viewing not only adaptive anatomy, technology, and speech, but also the writings of Merleau-Ponty.

For Merleau-Ponty, one's desire to creatively delineate oneself, and self-show, continues to appear as the body's manifest activity even if, in the cases of anatomical or psychological trauma, the 'self' must entirely renew itself for this display. I introduced the notion of corporeal self-showing by articulating how, for Merleau-Ponty, animal bodies appear as corporeal communication. Such bodies and body organs like cell tissue, nerves, eyes, and hands, are already communication because in their interrogating (moving) and responding (perceiving), they signify to oneself and others their response. And I emphasized that although corporeity shows itself to exist *as* communication, it is not rooted in communication. On the contrary, communication sinks itself in and emerges from corporeity. I showed that, for Merleau-Ponty, animals—excluding the more simple animals—are "opened up to" their *Umwelt*. They move creatively in response to signs and intend-toward the creating and negotiating of their own bodily boundaries; they intend-toward, pointing to themselves and thus to others, too. With the support of research by Vallier and Sjöholm, I showed that, for Merleau-Ponty, animal behavior is constitutively communication rooted in corporeity that directs itself outward and makes of itself an appearance.

That corporeity exists *as* communication, and that such communication is rooted in the body is significant for several reasons. It shows the body as a whole displaying activity traditionally believed confined to the brain and cognition. In so doing it minimizes one of the problems inherited with Cartesian dualism: alienation from one's body and from nature. The assertion of a mind-body substance difference entails a deepening identification among humans, of 'mind,' 'idea,' or 'immateriality,' with their understanding of self. Yet, if Descartes's legacy shrinks the distance between human self-understanding and the 'divine' by attributing to the self qualities traditionally linked with the dominant European idea of God and compatible with Catholic doctrine about the soul's permanence after death, it does so for a price. Identifying with the idealist side of the Cartesian bifurcation installs as presupposition the absolute separation between one's apparent self and one's body and nature. By showing corporeity existing as communication—but that communication is of corporeity—my reading of Merleau-Ponty 'returns' communication (interrogation and response, signification, idea, and thought) to bodies and bodies to nature, nonreductively.

I have magnified this exposure of selves, bodies, and nature as interpenetrating, for Merleau-Ponty, through my examples of anatomical, technological, and linguistic adaptation. The example of spinal-cord injury shows how physiology can respond to physiologically grave twists of fate. In the case of high-level spinal-cord injury, we see the physiology adapting to its new conditions by forging new mechanisms for temperature control, and new rules for apportioning callused and live tissue—thus showing how the injured anatomy engages primordial forms of communication (interrogation and response), and fights to create new borders for the anatomical self. For Merleau-Ponty, this anatomical shifting shows bodily desire intending toward a self-showing that self and other signifies. With respect to technological adaptation, assistive technologies such as wheelchairs, walking sticks, and prosthetics extend and adapt the lived body in ways that anatomy cannot. Co-operating with anatomical adaptation, technological adaptation reshapes the surface of one's lived body as intending self-signification and self-display. And, through the example of traumatogenic therapeutic narrative, speech shows itself likewise to be *adaptive*. That is, speech negotiates its surrounding conditions of self and roots that negotiation not in the signifying of itself it shows, but in the desire that is prereflective corporeity. The basis of speech responding to traumatic experience appears to be the enormous demand of a prereflective violation. Judith Herman's second stage of trauma recovery shows itself not to be as Herman indicates: a proper representation of the traumatic episode, but rather, to be a *linguistic adaptation* of a preconscious wound that remains unacknowleged and unrepresented.

By bringing into dialogue recent research in spinal-cord injuries, trauma, and Merleau-Ponty studies, this essay discloses a framework showing adaptive anatomy, technology, and speech as interpenetrating. It provides new scientific evidence for Merleau-Ponty's view of humans and nature as co-conditioning and is virtually the only philosophical commentary to reveal and explicate Merleau-Ponty's implied concept of trauma.

Merleau-Ponty and Nietzsche on Embodiment

If for Merleau-Ponty trauma entails a certain impossibility of perspective, it does not obviate the fact that to be embodied is to live with a perspective. We might thus attribute to Merleau-Ponty a variant of "perspectivalism."[31] Living with a perspective for Merleau-Ponty presupposes, as was shown earlier, three aspects of embodiment—situational, mnemonic, and invisible. Merleau-Ponty's perspectivalism points to embodiment as its foundation. Moreover, like Nietzsche's, it is not only not ideationally reducible, it is also not materially reducible. For Merleau-Ponty, the three aspects make possible and co-arrange one's embodied experience. They point to embodied experience as involving idea and matter in such a way that the two are mixed inextricably. This is especially visible when one thinks about the idea of muscle memory implied in the hypothetical fantasy of Mary Lou Retton's mnemonic embodiment suddenly inhabiting John Malcovich's embodied situation. Her muscle memory includes certain types of learned gymnastic stunts and routines. We can imagine that for Merleau-Ponty, the fantasy of Mary Lou Retton's memory inhabiting John Malkovich's embodied situation would allow Malkovich's body to execute the gymnastics better than it otherwise would, if simultaneously putting his embodied situation in serious danger. What is significant is the implied mutual inextricability of so-called idea and matter exhibited in the example of muscle memory. This is important because it distinguishes Merleau-Ponty's work from strains of cultural theory, that is, certain receptions of Friedrich Nietzsche, Michel Foucault, and Judith Butler, for instance, which reduce their genealogies of bodily being to cultural conditioning, and cultural conditioning to ideas and ideational relativism. Even if body exists as communication for Merleau-Ponty, such communication does not reduce to idea. Nor does the body exist *of* communication. Merleau-Ponty's position thus also sets itself apart from those, which at their logical extreme, reduce perceived embodiment to language, idea, or spirit (G. W. F. Hegel (1807/1979), Jacques Derrida (1967/1974), Ernesto Laclau (1996)).

Merleau-Ponty's foundational body has in common with Nietzche's (non)foundationally privileged body, that it also mixes materiality and immateriality without falling into a Cartesian dualism or a reductive idealism or materialism. Such mixing is apparent for Merleau-Ponty as described above via situational, mnemonic, and invisible embodiment. It is also apparent in several additional ways, each of which invites comparison with Nietzsche. We saw how Merleau-Ponty's idea of intentional arc shows itself in the examples of verbal, anatomical, and technological adaptation. These adaptive behaviors co-express, -interrogate, and -respond to bodily desire. As interpenetrating co-constituents of bodily desire, alteration to one adaptive process effects new circumstances for the others. This relational dynamic is reminiscent of Nietzsche's dynamic nondualism that I have described as a field of co-operating planes, especially the planes of self, conscience, and corporeal punishment.

Merleau-Ponty's thought compares to other relational aspects of Nietzsche's thought. These indicate that determinate identity is probably a fiction—that rather than *relata* there appear to be *relations*. The lived body, for Merleau-Ponty, experiences by comparison 1) horizonal aspects of perception (i.e., boundaries of perceptions interrelated with the enveloping surroundings); and 2) arising aspects of perception (i.e., perception beginning-again and free of preestablished rational order). Such horizonal ongoing "beginnings" show themselves conditioned by three aspects of body: as situation, as memory, and as invisibility. As such, perceptions as horizonal and arising respectively, underline an indeterminacy and relationality to lived experience. Perceptions appear not as reifible things but imbedding relations. Thus Merleau-Ponty's project, akin to Nietzsche's in these respects, indicates that a so-called "experience" of a determinate identity is probably a deception.

The idea of a reified entity has traditionally presupposed a proper, conceptually articulable definition. For Nietzsche, as for Merleau-Ponty, the possibility of experiencing things as reified would have to begin with bodily perception. Perceptions as perceptions are in themselves, for Nietzsche, translations—designing the forces they interpret. They are thus metaphors, writes Nietzsche in "On Truth and Lying in a Non-Moral Sense" (TL 139–153), implying the Aristotelian distinction between logically proper analogy (concepts) and logically improper analogy (metaphor). Objects as perceived for Nietzsche, are necessarily "improper"; they are not logically proportionate to the external world forces they interpret.

We are reminded here of Merleau-Ponty's view that being exists as perceived, that is, as a perceiver that experiences oneself as perceived. And, being necessarily involves interpretation by the perceiver. Being as perceived, points to the impossibility of proper objects or concepts. A perceiver not only twists via translation the forces of that perceived, but the twisting is done toward and for the purpose of defining and pointing to oneself—and doing so for oneself and for presentation to others—that is, for other perceivers. Thus, Merleau-Ponty's notion, that being exists as perceived, also implies a soft teleology that determines his philosophy of nature moreso than Nietzsche determines his. If both Nietzsche and Merleau-Ponty underline the so-called indeterminacy and "impropriety"[32] of objects, Merleau-Ponty does so with a view of corporeal nature that is more teleological and determinate.

Nietzsche and Merleau-Ponty's views of corporeality regarding opacity invite parallelism, too. For Nietzsche, opacity of the body locates itself in his view of the body existing as and through perception understood as metaphor. The body shows itself to Nietzsche to be a (non)foundational foundation—an infinite depth of metaphors or surfaces[33] that suggest that the lived body itself not only exists as and through the "improper" metaphor, but necessarily shows itself to itself (even as improper) improperly. Merleau-Ponty's view of embodiment as opaque situates the opacity within a more systematic theory of perception. It decisively locates the opacity of the body according to a certain conditioning mode of embodiment and distinguishes this from other modes that are, if not properly graspable, at least perceptible. For Merleau-Ponty, there is not as for Nietzsche, the consideration that the notion of embodiment as both visible and invisible is itself also a deception. Hence, for Merleau-Ponty, corporeality appears to be a foundation of experience and not as for Nietzsche a (non)foundational foundation.

This is no small difference between Nietzsche and Merleau-Ponty, and it parallels differences in their styles of argumentation. Merleau-Ponty's relatively systematic formulation of the opaque body and decisive appointment of the body as basis of experience is reflected in the stark difference of his rhetorical style. Nietzsche's style is literary, poetic, exceeding linear argument; Merleau-Ponty's is scientific, empirical, and inscribed within more traditional lines of reasoning. Nietzsche's flamboyant, playful desystematization invites and anticipates the overcoming of the outermost limits to which his thought points. By contrast, Merleau-Ponty's more traditional argumentation builds into his position a more restrained approach to limit. It does not beckon to or prevision its own overturning. It points rather to the inevitability of newly and differently perceived particulars underlining it. With respect to system and limit, Nietzsche's reflections are the more radical.

Certainly both Nietzsche and Merleau-Ponty have an audience in the twenty-first century, but the views of neither reflect the common-sense views of contemporary Westerners. Speech and thought among contemporaries in the West remain infused not only with reflexive, subject-object and self-other machinery, but with modern interpretations of them. These generally assume the determinacy of the object and its distinct separateness from both other objects and the subject. This bent of contemporary modernity generally privileges a substance metaphysics (Scholasticism, theism, deism, Cartesianism) entailing substantial creatures and supreme substance. At the same time, the perceived problem of estrangment—from prime substance, God, nature, and bodies (one's own as well as others') seems also for many a common experience.

That Nietzsche and Merleau-Ponty have had to strain to make visible their respective ontologies of relatedness—of self and other, idea and matter, mind and body—indicates how entrenched modern and counter-modern[34] speech is in a substance oriented metaphysics. David Abram suggests that the primary origin of this entrenchment rests in the rise of phonetic speech.[35] The speech of Heraclitus, perhaps Nietzsche's favorite philosopher, occurs during the transition in the West to phonetic speech. An analysis of Heraclitus's thought is therefore warranted. The following discussion considers Heraclitus's thought as a harbor of prereflexive thought structures, and the possible significance of such structures for relational ontologies—like the (non)foundational and foundational relational ontologies that I attribute to Nietzsche and Merleau-Ponty respectively.

CHAPTER 8

Nietzsche before Nietzsche

Heraclitus's Speech Opening Nietzsche's and Ours to Preliterate Perceptual Structures

PART I. THE PROMISE OF HERACLITUS'S SPEECH

Broadly absent from human experience today is an ability to perceive a sensuous world exceeding a realm of human technology.[1] Our speech is a human technology—structured by and for humans in ways it has not always been. How might the ways humans perceived prior to technological speech differ from how many of us do today? By "technological speech" I mean speech coinciding with the spread of phonetic writing[2] and literacy.[3] Such a spread occurred in the West roughly around the time of Socrates.

Do we perceive landscape, living plants, and animals—realms arguably beyond our technological realm—differently than humans did prior to the spread phonetic writing? If our perception is different than that of our pre-Socratic predecessors, how might this insight foster an interpretation for Nietzsche's cryptic reference to Heraclitus in *Twilight of the Idols* (1889/1968b)?

I will not offer a scholarly exegesis of Nietzsche's Heractlitus,[4] but rather a reading of Nietzsche and Heraclitus in the 'spirit' of Nietzsche's own philology. Such a spirit, MirkoWischke (2002) notes, is the opposite of the "encyclopaedic science" into which, to Nietzsche's dismay, ninetheenth-century philological studies had morphed[5] (99). Philology is not meaningful, for Nietzsche, as an end in itself but as a discipline "acting counter to our time and thereby acting on our time and, let us hope, for the benefit of a time to come" (UDH, Foreword).

Nietzsche's appeal to Heraclitus's speech in *Twilight of the Idols* is an appeal to a radically different perceptual existence. It alludes to resources, in Heraclitean thought, for expanding and interrupting habitual perceptual structures that Nietzsche's own nineteenth-century European speech had inherited. For the Nietzsche of *Twilight of the Idols*, certain structures of Heraclitean speech signify a promise for invigorating life. Via opportunities latent in its own obscured pre-Socratic/prephonetic ancestry, technological speech can open up and transform the parameters of its own being.

My analysis of the promise of Heraclitean speech begins with the idea of a shift in human perceptual structures in the West. The shift coincides with the transition to literacy between the seventh and fourth centuries B.C.E. Several attributes, generally distinct from the perceptual structures of literate speakers, can be said to manifest in the speech of preliterates, such as Homer, and transitional figures such as Heraclitus. I focus here on analyzing three such preliterate perceptual structures. First, prephonetic writing can be shown to explicitly depend on a "more-than-human" (Abram 1996) sensuous world. With the advent phonetic writing, by contrast, script no longer depicts images of the intended object, as in pictographs, but instead *sounds* of the word that signify the object. Such a sound system—phonetic speech—becomes increasingly independent of the sensuous world of creatures and geography that otherwise had been a requisite participant in the flow of human communication. Phonetic speech seems no longer to require the participation of a "more-than-human world."[6] It happens, as David Abram argues, via a technology created especially by humans, for humans (1996, 93–136). It appears to require little participation from creatures and beings that are not human. With this new technology, human communication and thinking can be said to undergo a perceptual shift away from a more-than-human world (Abram 1996, 93–136). Second, an antiquated verb-tense called "the middle voice"—basically extinct in modern Western languages—is harbored in Heraclitus's speech. The middle voice is noteworthy in its ability to structure perception independently of a subject-object construction. Such a perceptual structure seems all but inaccessible to us today, for whom a subject-object logic can be said to permeate and delimit our perception as much as our sentences. Third, the significance of *psyche* (soul), in Heraclitus's fragments, differs in critical ways from both its preconceptual Homeric predecessors and conceptual Socratic successors. *Psyche* has not yet come to signify a persisting, identifiable, rational concept associated—according to a dominant reception of Plato in the West—with human nature.

Before I begin my analysis of the promise of pre-Socratic speech—and especially Heraclitean speech—for invigorating life, it will be helpful to review here the context in *Twilight of the Idols* in which Nietzsche praises Heraclitus. In it much life and death imagery appears. Nietzsche links his references to Heraclitus to the "life" side of this imagery. "They believe that they do honor to a thing by divorcing it from history *sub specie aeterni*—when they make a mummy of it. All that philosophers have treated for thousands of years, have been concept-mummies; nothing real has ever come out of their hands alive" (KSA 6, 74). Nietzsche criticizes philosophers, accusing them of isolating entities from the complex forces and contexts of their environment. This, he writes, is like Egyptian practices of mummifying dead kings. Philosophers do so, however, with ideas, not kings. They wrap and bind them, delivering them dead, which is to say, in the form of an abstract concept. Intact in another time and place the concept arrives like mummy—extracted from the situational particulars that made it live. If partially preserved, it is not at all 'alive.'

The text's first chapter, "The Problem of Socrates," opens with an image of a raven circling in on the sight and smell of a dead carcass. The death imagery, here in Nietzsche's first aphorism, arises with a discussion about wisdom. The wisest humans throughout the ages, says Nietzsche, have always come to the same judgment about life: "It is good for nothing" (KSA 6, 65). "Even Socrates' dying words were:—'To live—means to be ill a long while'" (KSA 6, 67). "[T]here must be some sickness here," says Nietzsche (KSA 6, 67). Wisdom and philosophers traditionally associated with wisdom, appear to be life enervating. Socrates and Plato are "symptoms of decline" and "instruments in the disintegration of Hellas" (KSA 6, 68). Nietzsche for the most part sustains this critique against philosophers in general and Socrates in particular throughout the first two aphorisms of *Twilight of the Idols*. There is, however, one glaring exception: Nietzsche's critique of Heraclitus.

> I set apart with high reverence the name of *Heraclitus*. When the rest of the philosopher crowd rejected the evidence of the senses because these showed plurality and change, he rejected their evidence because they showed things as if they possessed duration and unity. Heraclitus too was unjust to the senses, which lie neither in the way the Eleatics believe nor as he believed—they do not lie at all. It is what we *make* of their evidence that first introduces a lie into it, for example the lie of unity, the lie of materiality, of substance, of duration. . . . 'Reason' is the cause of our falsification of the evidence of the senses. In so far as the senses show becoming, passing away, change, they do not lie. . . . But Heraclitus will always be right in this, that being is an empty fiction. The 'apparent' world is the only one: the 'real' world has only been *lyingly added.* (TI, "'Reason' in Philosophy"[7] 2)

Heraclitus is the only philosopher Nietzsche exempts from his otherwise relentless criticism of philosophers. Nietzsche gives us only general reasoning for this singling out: Heraclitus rejects the belief that things have "unity," "substance," and "duration;" and says Nietzsche, "Heraclitus will always be right in this, that being is an empty fiction" (TI, RP 2).

In "'Reason' in Philosophy," Nietzsche blames the rise of a certain practice of reason for falsifying the testimony of the senses. We might assume, therefore, that Heraclitus's thought precedes such reasoning habits. But what are these habits? And what would thought prior to them be like? What forces amidst the habits of a pre-Socratic speech become largely absent in technological speech? In this chapter, I shall bring some of these obscured forces into view. By excavating Nietzsche's undeveloped praise for Heraclitus, I will disclose the promise of Heraclitean speech for opening up Nietzsche's and our own.

PART II. THE SHIFT FROM PRELITERACY TO LITERACY

Among scholars, the Heraclitean fragments continue to elicit variegated and conflicting interpretations. One reason for a particular range of the interpretation that arises in the twentieth century is that Heraclitus speaks a speech in transition. Between the sixth and fourth centuries B.C.E, the Greeks underwent a transition from an oral to a literate culture (Havelock 1982, 1983). By considering this shift, one sees some[8] of the reasons why Heraclitus's fragments have evoked such varied interpretations—why, for instance, some have interpreted Heraclitus's speech as an organized system with fire at its center (Vlastos 1955), and others have seen it as a complex of homonymous meanings exceeding coherent synthesis (Waugh 1991). In either case, the fact remains that Heraclitus's sayings are spoken to an audience in transition *to* literacy. Heraclitus's speech can be richly interpreted when taking the fact of this development into account.

Eric Havelock (1982) shows that as recently as two generations ago, three unexamined assumptions often guided the way Western scholars read and taught Greek studies (220–223). These assumptions are still influential today. First, classicists viewed the Greek culture from Homer to Aeschylus to Aristotle, in other words from the eighth to the fourth centuries B.C.E., as an entirely literate culture. Second, they engaged the Greek language as though it were a system of interchangeable signs conveying meanings that operate as a constant. By interchangeable signs, I mean that that which, say, *logos* denotes for Hesiod agrees with that which *logos* found in a similar context[9] signifies for Plato. Third, scholars assumed that if language mirrors thought and the Greek language is a system governed by a code consistently linking a word in a given context to a set meaning, then Greek thought also resembles such unity. This does not mean the absurdity that Homer's thought could be reduced to Thucidides's, but that Homer could converse with Thucidides using the same vocabulary and both would generally understand each other. In the language of Saussure (1959) their words would bring to mind the same "signified."

By the second half of the twentieth century, however, some interpreters began questioning these presumptions (Havelock 1982, 224–225). Although the old assumptions still influenced the way some people read eighth-to-fourth-century-B.C.E. Greek literature, many interpreters began to acknowledge that poets of archaic Greece (Homer) and Attic Greece (Heraclitus, Aeschylus, Sophocles) lived amid and created for an entirely (archaic Greece) or predominately (Attic Greece) oral tradition. Havelock cites two crucial works that taken together, indicate the start of the transition to a literate culture, to be no earlier than 720 B.C.E, and places the date of the Homeric epics before this time. In 1928, Milman Parry showed the latter on the basis that such verse constitutes an oral instrument that can be comprehended only under the assumption that it was created and performed by singers who could neither read nor write (Parry 1928, 1971). This coupled with Rhys Carpenter's work (1933, 8–29; 1938, 58–69) in the 1930s indicating convincingly that the Greek alphabet could not have been invented earlier than 720 B.C.E., locates the beginning of a transition, from an oral to a literate culture, at about this time. If one focuses on the assumption that early Greeks lived an oral tradition, how might this assumption alter the way one of a literate culture reads the texts? If the words of Hesiod and Homer, and of the more transitional figures like Heraclitus, cannot be understood with the same interpretive methods as are used for the writings of Plato and Aristotle—how are we to interpret them? Moreover, were we to find a credible approach, what sorts of perceptual differences would we find in our preliteracy heritage?

PART III. RISING TECHNOLOGY: THE ALPHABET AND PHONETIC SCRIPT

One way of accessing the earlier speech is to excavate the rise of the alphabet in the West—the building blocks of phonetic script. Research in the development of writing shows that virtually all of the early human writing systems rely upon a more-than-human-world: local animal life, vegetation, seasonal changes, and landscape. Pictograms, pictures intended to be "read," are generally elements of most earlier writing systems. Early Sumerian pictograms from circa 3000 B.C.E., overflow with stylized images of sunrises and birds, cows and oxes, as well as waves and water wells (Robinson 1995, 50). The stylized Sumerian images of the bird and ox are pictograms. The image stands "literally" for the word intended. The stylized Sumerian images of the sunrise and the waves however are called "ideograms"—ideograms often draw an image closely associated with the intended idea. The sunrise expresses the idea: "day." The curvy lines or "waves" communicate the associated idea: water. "[T]he gliphs which constitute the bulk of these ancient scripts continually remind the reading body of its inherence in a more-than-human field of meanings" (Abram 1996, 97).

The path from pictographic and ideographic writing to phonetic script may be explained by expanding the use of pictographs to be read phonetically, not pictorially. Such symbols are called "rebuses" (Robinson 1995, 42). Certain basic words that do not easily correspond to a sensory image, for instance, the proper name of the Egyptian king, "Ramses," is difficult to put in a literal pictogram. It could with relative ease, however, be communicated by drawing pictures to be read not literally but phonetically—in terms of the sound they bring to mind. In order to write the first part of "Ramses," Egpytian hieroglyphs resort to an image of a sun. The word for sun was pronounced "R(e)." In order to invoke the sound "R" the sun was drawn. It marked the first of several rebuses spelling "Ramses" (Robinson 1995, 42). Almost any otherwise difficult-to-draw idea—imagine drawing a literal pictogram for the word "idea" for instance—could be "spelled" via such pictographic puns (Abram 1996, 98).

With a little imagination, one can see how pictograms eventually lead to phonetic script. "In the ancient Middle East the rebus system was eventually generalized . . . to cover all the common sounds of a given language. Thus, "syllabaries" appeared, wherein every basic sound-syllable of the language had its own conventional notation or written character" (Abram 1996, 99). These notations often had their origin in a rebus. A notation system that would eventually give rise to the alphabet was "itself developed by Semitic scribes around 1500 B.C.E." (99). These notations reflect the innovative system for paring down to fewer images, the basic notations needed to sound them. Integral to this system was a recognizing of the dual role of consonants and sounded breath (vowels) in speech. "The silent consonants provided, as it were, the bodily framework or shape through which the sounded breath must flow. The original Semitic *aleph-beth*, then, established a character, or letter, for each of the consonants of the language" (99).

Even if we generally do not experience a world of sensuous beyond-human-world phenomena in our speech or alphabet—as earlier speakers arguably did—we can trace in our present language such experience. For instance, the first letter of the English alphabet, "A" comes from the first letter of the early Semitic *aleph-beth* (alphabet) written "A" but turned 120 degrees counterclockwise—that is, it is almost an upside-down "A" (Abram 1996, 101). An "A" turned almost upside-down draws the early Semitic pictogram for *aleph* (ox) (Abram 1996, 101). It represents an ox head and horns. Thus, the genealogy of the first letter of the English alphabet harbors this early phonetic heritage: a pictogram to be read not pictorially but phonetically—as the sounded breath that begins the Hebrew word "aleph" (ox) (Abram 1996, 101). Indeed, when speakers of English write and speak the letter "A" today, we are entirely unaware of the stylized image of an ox that at one time was a participant-by-association in one's experience of reading and writing words with an *aleph* sound.

With the emergence of phonetic script, writing not only stops requiring readers and writers to experience such sensuous images when speaking certain words, but also appears to stop requiring one's dependence upon a more-than-human world. The human activity of speaking, writing, and reading, begins to seem as if independent of a more-than-human world that earlier speech explicitly depended on. With the advent of such writing "a new distance opens between human culture and the rest of nature" (Abram 1996, 100). If pictographic writing initiates this distance by first displacing our "sensory participation from the depths of the animate environment to the flat surface of our walls, or clay tablets, or the sheet of papyrus" (Abram 1996, 100), then alphabetic writing, it seems, distances us even further. Alphabetic writing radically shifts the context of our perception. Our perceptual practices had come to dispense with the sensory phenomena that surrounded us in nature, practices that had been overtly sensible in the preliterate modes of writing, speaking, and living of our ancestors.

When Homer recorded, in a phonetic alphabet, the oral rhapsodic stories of his ancestors, the information lodged in the epic tales appeared for readers visibly stationary. It rested in a form that could be contemplated, set down, walked away from, and returned to later for further questioning and examining. As Eric Havelock (1986) sees it, "It is only as language is written down that it becomes possible to think about it. The acoustic medium, being incapable of visualization, did not achieve recognition as a phenomenon wholly separable from the person who used it. But in the alphabetized document the medium became objectified" (112). With this new phonetic writing comes a powerful force for shifting speech and perception in the West. Human speech would develop patterns that repeat and underscore a subject-object grammar. Literacy fosters this development because literate humans are more likely to experience themselves as perceiving subjects, and sensual phenomena (i.e., script on a page) as objects that are identifiably distinct and controllable by humans. The advent of phonetic writing gives rise to a perceptual shift, in which humans see themselves, more strongly than ever before, as authors and agents of actions and consequences. Just as words and their referents appear to be stable objects—frozen as well as infinitely manipulatable according to a human agent's whim—so too does the human concept of self reflection appear to be the structure of this new scribing technology.[10] Human beings' perceptual experience in general could be said to shift according to new limits and habits of speech. With the rise of literacy, human speech, and self-understanding begin to appear as if independent from a sensuous "more-than human world" of local vegetation, neighboring landscape and resident creatures.[11]

Twenty-first-century philosophers of mind, language, and cognition, generally focus on processes of contemporary speech and perception. These studies do so with little concern about the possibility that the structures permeating contemporary speech and perceptual processes may not be necessary, nor exhaust pertinent structures of such processes. Classicists and philosophers, likewise, generally neglect implications of the rise phonetic speech in the West (Abram 1996, 123).

> Most of the major research, in other words, has focused upon the alphabet's impact on processes either internal to human society or presumably "internal" to the human mind. Yet the limitation of such research—its restriction within the bounds of human social interaction and personal interiority—itself reflects and anthropocentric bias wholly endemic to alphabetic culture. In the absence of phonetic literacy, neither society, nor language, nor even the experience of "thought" or consciousness, can be pondered in isolation from the multiple nonhuman shapes and powers that lend their influence to all our activities. (Abram 1996, 123)

In comparison to the genealogically prior pictograms and ideograms, phonetic script as visual image has almost no visually sensual information supporting the meaning that the script intends. Analysis of speech prior to the transition to phonetic literacy allows us the chance to see the limits of our own speech and perceptual structures. This seeing shows perceptual limits to be greatly conditioned, by human cultural and historical developments. More important, it suggests the existence of expansive modes of human speech and perception other than our own—modes that have remained for millennia unimagined and thus unexplored.

Just as today many would be surprised to learn about such possible expanses, so too during the Spanish conquest of Mexico, the Aztecs were awed by the Spaniards' ability to read phonetically. It seemed magical. According to Tzvetan Todorov (1984), it is precisely the Spaniard's technology of phonetic script that enabled the lightening fast conquest of Mexico by Cortéz and his few-hundred men over Montezuma and his several thousand (Abram 1996, 133–135). The writing of the Aztecs, like most other prephonetic scripts, was replete with pictograms and ideograms. It was very much engaged in a more-than-human environment. The perceptual structures of the Aztecs were delimited by the construction of their speech and writing. The Aztec manner of speaking and writing, in turn, depended on a sensuous world of earth, air, plants, and animals. "Everything happens as if, for the Aztecs, [written] signs automatically and necessarily proceed from the world they designate" (Todorov 1984, qtd. in Abram 1996, 134). The phonetic script of the Spaniards, by contrast, does not implicitly refer to a more-than-human environment, but instead refers back to themselves. It refers back to a human technology that can communicate the sounds of the intended words while circumventing pictographic imagery and more generally, the extra-human realm on which prephonetic thought and communication depends. The speech of the Spaniards, abstracted from the more-than-human sensory world, shifts the perception of the Spaniards away from a more-than-human environment allowing them to communicate primarily in a human realm.

More important to Todorov's theory is that such self-referential speech, independent of the surrounding sensual situation, seemed to the Aztecs like magic. The Spaniards seemed to possess direct speech of the gods, independent of landscape, seas, and other creatures. They could read from books or their own notes as if to suggest that the "leafs" are talking to them, "for the black marks on the flat, leaflike pages seemed to talk directly to the one who knew their secret" (Abram 1996, 132). Todorov shows that the Aztecs not only found themselves at a strategic disadvantage, because their speech and writing could not veil their intentions, but more important, their speech placed them at an overwhelmingly psychological disadvantage. The Aztecs' words and script appeared to belong more to the sensuous more-than-human environment around them than to themselves. "To be duplicitous with signs would be, for the Aztecs, to go against the order of nature, against the encompassing speech of logos of an animate world, in which their own tribal discourse was embedded" (Abram 1996, 134). By contrast, the speech of the Spaniards, by virtue of its apparent independence from the beyond-human world, communicates almost exclusively to and via humans and human-made objects.

Comparable to the way the Aztecs experienced their speech and script—connected with the pulsing and mysterious creatures reciprocating life to the Aztecs, are certain ways the Hebrew scribes, especially prior to phonetic script, experienced the letters of the *aleph-beth*—as living.[12] Have humans, living in the West since the fourth century B.C.E., generally lost such a perception of living speech? And if so, what other human perceptual experiences remain occluded from our experience since the rise of literacy?

Prior to the rise of phonetic script, the idea of "psyche" and other ideas, such as "justice," "rationality," and "goodness," were arguably experienced as inseparable from the events among which they were orally said and acoustically heard.[13] With the spread of phonetic writing, such ideas take on the appearance of autonomous subsistence or of having originated from a realm of such being (i.e., Plato's forms). As I have shown in chapters 2 through 6, the writing of Nietzsche interrogates the prevailing belief in the West—a belief that emerges with phonetic writing—that humans can properly conceptualize themselves and perceptual objects. Thus, Nietzsche's nod toward Heraclitus's transitional speech—a speaking not yet spoken to a phonetically literate audience—registers yet another gesture by Nietzsche toward expanding his structures of experience beyond their habitual limits. Indeed as I have suggested here, the orientation of speech for the preliterate speaker and listener appears foreign to his and our own. Our perceptual structures are rendered virtually powerless if asked to perceive beyond our usual subject-object, cause-and-effect patterns. In other words, we typically do not and cannot[14] live a speech and nature that coordinate one another and obscure boundaries of selves and concepts. It is not overstating the matter to say that as we speak and think we typically do not perceive a sensuous natural world supporting that very reflection, whereas preliterate humans did; moreover, that we have an exceedingly difficult time even imagining what such perception would be like.

PART IV. ANCIENT SENTENCE STRUCTURE: A VERB TENSE WITHOUT SUBJECT OR OBJECT

There is at least one way that we can begin to imagine the terrain of preliterate speech and perception: by way of an antiquated Greek verb tense known as the middle voice. The Greek middle voice descends from a verb tense in early Sanskrit that exists only in a truncated reflexive form (i.e., "let it become strong") in present Western languages (Scott 1990, 19).[15] It describes an event that occurs neither as activity nor passivity but suspended between these. The identity of subject and object in such an event becomes obscure and the speech itself unfurls in a manner that can be called nonreflexive, that is, devoid of a subject-object formation. Although the middle voice operates in either a transitive or intransitive mode, it is the latter that most concerns us here. This is because the intransitive middle-voice verb does not take an object and thus avoids inscribing thought within a reflexive structure. Its significance lies in its ability to point to a way of speaking and perceiving not circumscribed by a subject-object relation.

Nevertheless, let us begin with an example of the reflexive middle voice so that we can subsequently contrast it with the nonreflexive middle voice. This shows itself in "pacate," a transitive[16] middle voice verb of early Sanskrit. It can be rendered in English "(the cook) cooks for himself" (Scott 1990, 19), and describes a subject that is implicated in the result of the action and yet "is neither the active subject nor the passive object of the action (19). The latter suggests reflexivity via the word "himself." In such transitive middle voice modes, the fullness and ambiguity imbued in the middle voice expresses itself via a self-relational structure. It is precisely this self-reverting structure that is absent in the intransitive middle-voice verbs. For instance, the intransitive verb "drmhate" means "(something) becomes firm" or "firming comes of its own action" (19).

One can imagine intransitive middle-voice verbs describing the activity of "grass," "cleaning," and "discipline." In English one might portray these as "grass grasses," "cleaning cleans," and "discipline disciplines." Each of these implies carries an action born of the subject, but rather than affecting an object separate from it, or reflecting the action back upon itself, each leaves the consequence of the enactment open-ended, thus evoking a sense of enigma, overflow, and ambiguity.

Our own way of speaking describes this middle-voice activity in terms of the action somehow returning to the subject. Because of this it still involves "reflexivity"—a structure of thought beyond which we find it so difficult to think (Hansen Quinn 1987, 163). Yet, if one begins with the idea "fruit fruits," or more specifically, "blackberry blackberries," and tries to rid this notion of any admixture of reflexivity, one can begin to see the blackberry action as activity whose source defies any single origin or agency. The "blackberrying" activity, rather, arises among and cooperates across interpenetrating nature. The sources and the destinations of the blackberrying become indistinguishable from one another as others. The subject—the blackberry—traditionally perceived to be enacting the action, becomes part of a sweeping movement. The forces born of this movement (i.e., blackberry seeds, bushes, and berries) and those neighboring it (i.e., soil, sunshine, moisture, minerals, and rainwater) gather and take part in the broad indeterminacy of the action itself. The subject appears to be the same as the movement it enacts, and in an ambiguous way, suggests an inclusion of the environmental forces in its immediate sphere. These neighboring forces reveal themselves as participants in the interplay of forces enacting and constituting the movement itself. In this way, a sense of overflow emerges as not only the subject-object formation disappears, but also as the boundaries delimiting the subject recede. Any such limits reveal themselves to be highly transitory, rendering the so-called "subject" coextensive with the range of forces permeating the activity. It is this kind of expansive, prereflexive expression that characterizes the middle voice.

In a similar way that the early Sanskrit middle voice—especially its intransitive middle voice—displays a structure of indeterminacy and prereflexivity, the middle voice in the ancient Greek language does too. The common characteristic of the Greek middle verbs reveals that the subject of the verb does the action, but that the subject has an interest in the act that somehow returns to the subject (Hansen & Quinn 1987, 163). Generally, the meaning of a Greek middle verb renders itself as "doing something for oneself" (316). Yet various competing ways of translating middle-voice verbs arise, depending upon the verb and the context.

Middle deponent verbs lack forms in the active voice, but middle deponents are translated by an English active form, for example, *δέχομαι* (I receive). The future middle of certain verbs bears a passive meaning: *ἕξομαι* (shall be held). Some verbs in the middle voice have particular meanings that must be memorized, yet these often show the subject having some sort of special interest in the action. We can compare the respective active and middle voices of the following three verbs: 1) active voice: *αποδίδωμι* (give away); middle voice: *αποδιδωμαι* (sell); 2) active voice: *παύω* (make stop); middle voice: *παύομαι* (cease, stop); 3) active voice: *τίθημι* (to make, that is, a king makes a law for his people); middle voice: *τίθεμαι* (the people of a democracy make a law) (744). If we focus on the above examples of middle-voice verbs, we see that the action returns to or interests the subject in the following ways. When one sells something, one is paid in return; when a person or thing ceases, it stops itself and not something else; when a law is made by the people of a democracy it applies to the people, and not some body of forces separate from those which enact it.

By enumerating the various uses and functions of middle verbs one sees that the Greek middle verbs bear a middle-voice form but an active or passive meaning. All of the above examples are transitive middle verbs indicating subject-object reflection. For instance, the first two of the three examples in which the middle voice bears a passive meaning or an active meaning respectively carry a reflexive signification. In the third example, if people in a democracy are the "subject" and they make a law for themselves (the "object"), the reflexivity is present. Similarly, if a king imposes a law upon others, that is, his or her subjects, the use of the word still carries a reflexive meaning.

Nevertheless, the combination of the reflexive, semantic significance and the middle-voice form can be said to carry, and cover over, a nonreflexive significance often characteristic of the middle voice. In our discovering that which in the middle-voice heritage is covered over, we can begin to see these Greek middle verbs suggesting, by their form and lineage, an expansive ambiguous event implied by the nonreflexive middle-verb form of their Indo-European ancestry. Greek middle verbs generally announce their meaning according to a self-relation. The ancient Indo-European, especially Sanskrit, heritage of the Greek middle-voice form, however, implies a nonreflexive semantic significance harbored in this announcing. Viewed in this way, the self-relational meaning of Greek middle-voice verbs can be seen covering over an obscured nonereflexive significance. In the latter a "subject" and "object" do not emerge, but are dissolved in a pool of ambiguous, countervailing forces. "[T]he middle voice in these instances can indicate a whole occurrence's occurring as a whole without self-positing or reflexive movement throughout the event" (Scott 1990, 20).

Of the middle-voice verbs in the Heraclitus fragments, most are transitive. I have chosen one example of a transitive middle-voice and one of an intransitive middle-voice verb in fragments D. 101 and D. 129 respectively.[17] The transitive formation of D. 101 reads:

> *ἐδιζησάμην ἐμεωυτόν.* (D.101)
> I searched out myself. (Kirk & Raven #246)
> I went in search of myself. (Kahn #28)
> I have searched myself. (Wheelwright #8)[18]

Emerging from the aorist indicative middle *ἐδιζησάμην* is a specific sense that is consistent with the general middle-voice meaning of "doing something for oneself." The subject does the searching, and the enactment takes place within the sphere of the subject. "Myself" plays the role of object and the sense of movement originates from and returns to the subject. It is self-reflexive. Despite the overt reflexive structure and semantic significance of the phrase, the middle-voice form implies—by virtue of its heritage and that which it conceals—a nonreflexive occurrence of searching. The subject "I" bears both the imprint of movement it enacts self-relationally and an imprint that suggests another contour to the subject. This concealed nonreflexivity blurs the distinctions between the "I" and the action, and enacts the event in a movement whose sources and destinations exceed the limits drawn by a subject-object relation. It discloses the supposed "subject" and "object," instead, to be reciprocally interpenetrating.

The idea of excess described above, when considered in the context of a transitive middle-voiced phrase, loosens the bonds of the surface reflexive structure. It expands and deepens the implied limits and significance of the subject. The subject's reflexive meaning becomes exceeded and obscured by the buried, nonreflexive significance it bears. The implied exuberance or excess of the middle voice intensifies in the intransitive middle voice, which is without an object—and so, nonreflexive). In D. 129 the intransitive middle verb *ἐποιήσατο* (to fashion [of oneself], to create [of oneself]; made his own) is structured by and signifies such erasure of reflexive borders.

> Pythagoras son of Mnesarchus pursued inquiry further than all other men and, choosing what he liked from these compositions, made a wisdom of his own: much learning, artful knavery. (Kahn #25)

> Pythagoras, son of Mnesarchus, practiced enquiry beyond all other men and selecting these made them his own—wisdom, the learning of many things, artful knavery. (Kirk & Raven #256)

Although Charles Kahn, G. S. Kirk, and J. E. Raven use "made his own" to translate *ἐποιήσατο*, I prefer "fashioned of himself" or "created of himself." "Made his own" suggests the significance of the middle-voice as a reflexive structure only. It indicates someone acting upon something outside of one's sphere to make it "one's own." Such a translation, and its significance, basically excludes the buried, nonreflexive meaning a middle-voice construction entails. The other two translations, in contrast, distance themselves from a reflexive meaning. Without the subject-object formation, *ἐποιήσατο* does not specify its enactment for itself nor for something else. This means that unlike in D. 101, the subject does not become also an object or "other" to itself in the course of the action. The action co-arises from the subject and the influences surrounding the subject. Its purpose, therefore, remains something ambiguous within the sphere of the action. Thus, rather than naming a simple value and object of that value, *ἐποιήσατο* can be said to bring to the fore a complex of competing valuations, expressing a certain destruction cum co-creativty, and freedom from definition. This intransitive middle-voice movement develops ambiguously out of its own interplays and tensions without a self-relational structure, as is the case with a subject-object formation (Scott 1990, 18–35). Whereas the transitive middle voice, if suggesting nonreflexivity in what it conceals, still maintains a certain order in its self-relational form, the virtual freedom from reflexivity implied in the structure of the intransitive middle voice leaves the identities of the participants and consequences of the movement entirely unresolved.

The structural remains of the Greek nonreflexive middle voice in Heraclitus's speech are remnants of an earlier way of human speaking and being. They expose a mode of human thought prior to its organization around fixed concepts and subject-object formulae. Although preliterate speech includes reflexive forms, it is not yet entirely dominated by them. Thus, whether thinking in terms of an entire fragment, a single signifier in a fragment, or concept of a subject, soul, or self—in the context of preliteracy, none of these carries a stable, identifiable conceptual meaning.

If preliterate speech lacks coherent determinations, it abounds with forces influencing its meaning(s). Indeed, even without any awareness of the middle-voice aspects of the speech of Heraclitus, a plurality of possible significance appears in his speech. His is a speech replete with puns, irony, word-play, and ambiguity. A message to gather from it is not a simple one: *logos* communicates plurally.

Here are just a few Heraclitus fragments indicating his play with words, his humor, and his ability to evoke several directions of interpretation from select words:

D.81 Pythagoras was the prince of imposters. (K #26)
D.37 Swine delight in mire more than in clean water; chickens bathe in dust. (K #72)
D.9 Asses prefer garbage to gold. (K #71)
D.96 Corpses are more fit to be thrown out than dung. (R)
D.22 Those who seek gold dig up a great deal of dirt and find little. (R)

D.70 Human opinions are toys for children. (K #58)
D.97 . . . dogs bark at whomever they do not recognize. (R)
D. 87 A fool loves to get excited on any account. (K #60)
D.124 The fairest order in the world is a heap of random sweepings. (K #125)
D.11 All beasts are driven by blows. (K #76)
D.4 If happiness consisted in the pleasures of the body, we should call oxen happy whenever they come across bitter vetch to eat. (R)
D.42 Homer deserves to be expelled from the competition and beaten with a staff—and Archilochus too! (K #21)
D.48 The name of the bow [βίος] is life; its work is death. (K #79)

One of the most notable attributes of Heraclitean speech is its paradox, punning, riddling, and repartee. The wit and word play of the Greek, even if partly lost in the translation, still expresses in the English a complex range of serious humor.

The multiple significations of each of these fragments are disclosed in an obvious manner. For instance, fragment D.9 invokes a range of competing connotations. It reads, "Asses prefer garbage to gold" (K #71). Asses or donkeys have often been thought to be obtuse animals. With such a presumption, one reading goes: Asses are so dumb that they choose garbage over gold. The latter presupposes the value humans recognize in gold due to its worth in the marketplace or its beauty as a mineral. Most of us, however, can imagine, or perhaps know, of a person who has transformed their desire for the use of gold (i.e., money) and what it can purchase, into a desire for gold itself. The sort of person I am referring to does not have a limited desire for gold for its own sake or beauty, but an unlimited desire for it, fueled by an abyssal obsession. The gold could be said to provide such a person about as much good as it would an ass, for which garbage proves the more useful. The ass, therefore—which chooses the garbage—is the more practical. Bits and pieces of the refuse, that is, discarded food scraps, provide sustenance for the ass, whereas gold offers none.

Indeed, the fragment recalls Nietzsche's perspectivalism, the view that experience involves interpretation, and that truth according to one point of view becomes falsity from another (Waugh 1991, 615). The initial reading, presuming the stupidity of the ass, privileges a certain human perspective, which values gold for its trade value or for its own sake. From the ass's viewpoint, though, value could be said to rest more in the garbage, than the gold. The following Heraclitean fragments also evoke meanings that can be compared to a Nietzschean perspectivalism:

D. 61 The sea is the purest and foulest water; for fish drinkable and life-sustaining; for men undrinkable and deadly. (K #70)

D.60 The way up and down is one and the same. (K #103)
D.111 Disease makes health pleasant and good, hunger satiety, weariness rest. (R)
D.82 . . . the most handsome of apes is ugly in comparison with [a member of the] human race. (R)

These fragments emphasize both a being's perspective and the plural meanings communicated by a single phrase. Heraclitus's multiplicity of meaning and emphasis on viewpoint, are like the ancient Greek middle voice in at least one important way. They call attention to the conditioned state of entities, registering a bond between an entity's being and the context supporting it. Like the middle voice, Heraclitus's perspectivalism speaks connections across subjects and objects, selves and environment, not isolated entities abstracted from them. With its deliberate wordplay, and perspectivalism, Heraclitus's speech underscores the idea of contextual forces conditioning the experiences and appearances of beings. It displays a mode of human thought not yet constrained by strict subject-object formulae and apparent, if illusory, conceptual determinations.

PART V. CONCEPTUALIZING PSYCHE: A PRELITERATE VERSUS LITERATE CONCEPT OF PSYCHE

By becoming aware of the less constricted, preliterate mode of speaking and thinking that precedes phonetic speech in the West, one can begin to imagine the structure of one's own perceptual existence opening up. Such unlatching, and the awareness which fosters it, sheds light on Nietzsche's cryptic praise for Heraclitus in *Twilight of the Idols*. Heraclitus's thought is exceptional among philosophers, suggests Nietzsche. It promises to invigorate life. It promises this, implies Nietzsche, and all the while the thought of the dominant many continues pressing life flat. Details as to why the philosophical thought of Heraclitus might hold such promise, details that might actually make vivid Nietzsche's otherwise only general explanation—that Heraclitus sees being as flux and fiction—Nietzsche leaves, however, virtually unsaid. My analysis of the transition to literacy in the West is an attempt to provide such details. Thus far my analysis has made two observations. First, that human speech and perception move from structural dependence to *apparent* independence from a more-than-human world; second, that during the spread of literacy, speech fades out a middle-voice sentence structure that had obscured a subject-object distinction.

An additional attribute of speech on-the-way to literacy, is the development of the concept of *psyche*. With the rise of literacy, an apparently unified concept of *psyche* emerges out of a more expansive, less definite, preliterate meaning of the term. In what follows I will appeal to and interpret select Heraclitus fragments. My purpose is to show that the significance of the term "psyche" in these fragments differs in critical ways from that of both its preconceptual Homeric predecessors and conceptual Socratic successors. *Psyche* has not yet come to appear as a coherent concept. It has not yet come to name certain apparently persisting, identifiable attributes that would later be associated with the ideas of human nature, personality, and soul. The Socratic concept of soul accords with a dominant reception of Plato's writings in the West. The preliterate use of *psyche*, by contrast, is less determinate and more ambiguous. It expresses a range of significations that include human life, while also connecting with a more-than-human world.[19]

Below—divided into three categories[20]—are translations of almost all of the extant Heraclitus fragments that include the word "psyche." Rather than attempt a reading for each, I will interpret D.45 and indicate how the others, in light of this interpretation, can be said to distinguish Heraclitean *psyche* from Homeric and Socratic *psyche*.

I. *Psyche* and *Logos*[21]

The limits of psyche you will not find even if you set off on every route, so deep a logos does it have. (D45)

Psyche has a logos which augments itself. (D115)

II. *Psyche* and the Bodily Senses

Psychai employ the sense of smell in Hades. (D98)[22]

Those [Homeric] psychai sniff about in Hades! (D98)[23]

It is hard to fight against desire; for it purchases what it wants at the price of psyche. (D85)

Worthless witnesses to men are eyes and ears, if their psychai are barbarous. (D107)

III. *Psyche*, Wet and Dry

A man when he is drunk is led by a mere boy, having his psyche wet. (D117)

Dry psyche is most efficient and best. (D118)

To psychai it is death to become water; to water it is death to become earth. Yet from earth comes water; from water, psyche. (D36)

The signification of *psyche* changes from the time of Homer to that of Plato.[24] Out of an Homeric idea of impersonal *psyche* develops a Socratic concept of personal *psyche*. Scholars have often assigned three related notions to Homeric *psyche*: "life," "breath," and "shade."[25] Interestingly, the connotations of these do not invoke human personality. For Homer, cognition, affection, and appetite—three attributes that, since Plato, have typically been associated with personality—belong generally to *phren, noos*, and *thymos*—organ-like, palpable experiences of bodily existence. Other words, such as *ker, etor*, and *kardia*, allude to the "heart." *Psyche* in Homer, however, has no such link with cognition, the body, or its organs. Its Homeric senses arise almost exclusively in situations "concerning (1) 'life' (and so *psyche* could be risked or lost) and (2) breath/wraith, the life-breath whose function is solely to animate the human body in life and to depart it in death" (Robb 1986, 320). "Shade," for instance, denotes the immaterial shadow that departs an individual's body at death and wanders among all of the other shadows in Hades. No longer in possession of *noos, phren*, or *thymos*, a shade leads a rather sorry existence as a mere semblance of a human, most of whose personality, death has left behind. During a person's life, *psyche* breathes life into a person, but does little else (Robb 1986, 317). It does not refer to perceptive, cognitive, or appetitive capabilities, as would a subsequent concept of *psyche* emerging with the rise of literacy.

A number of scholars agree that it was primarily, but not exclusively, Socrates, whose discourse late in the fifth century, radically altered the meaning of *psyche.*[26] According to this view, the widely accepted meaning of *psyche* during Socrates' time continued to resemble that of Homer. Although there were other views about soul, such as the Pythagorean, Orphic, and Democritean, most of these were contemporary with Socrates and did not affect the mainstream use of the term. None of these, however, did as Socrates' appears to have done: link *psyche* together with thought and feeling (Robb 1986, 321).

To his contemporaries, Socrates' apparently unified[27] concept of *psyche* and its parts, that is, thinking and feeling, would have sounded odd. For not until after his influence do the Greeks begin to associate "a conscious, feeling, thinking, moral, autonomous self, the source in the human person of cognition and of moral decision, of personal feelings and of memory" to one's *psyche* (Robb 1986, 321). It is likewise during this transformation that a firmer conception of moral responsibility arises for the human agent. Humans begin to see themselves as moral selves, responsible for their actions, rather than as partial agents who, on account of the trickery or bad influence of the gods, are not fully responsible for their actions. Moreover, humans begin to see the state of their moral selves as a direct reflection of the condition of their *psyche*.

Earlier in the chapter, I suggest that phonetic writing in Greece did not spread until about the time of Plato. Indeed, Plato's speech about *psyche* is not only marked by its concept of soul that joins thought and appetites. It also hypothesizes a form of an ideal soul, transcending breath and embodied souls. Plato's hypothesis of forms, some believe, depends upon the historical event of phonetic inscription (Abram 1996, 109–115). With the onset of phonetic writing, *logos* in the sense of "writing" can be said to reify itself.[28] Plato's forms, for instance, are said to write themselves into the memory of an immortal soul (*psyche*). This scriptural metaphor arguably arose only after the onset of phonetic writing itself. Comparatively stable referents of the new system of inscription would appear as objects for contemplation that one can return to again and again, regardless of how often their details slip from one's natural memory. Phonetic writing allows for a method of remembering that becomes independent of the individual memories, and detailed contexts, in which they occur.

Such independence is significant, as Abram notes, because it adds fuel to an emerging mode of human orientation away from certain "more-than-human" contexts of the natural landscape, air, animal, and vegetative life, toward an increasingly cerebral human-oriented way of existing (1996). When one reads phonetic writing, one is primarily human-oriented in part because the medium is an abstraction from the sensuous details of nature once perceivable in earlier forms of writing. The dependence of phonetic writing on those details becomes all but invisible. The practice of phonetic reading, for instance, underlines instead communication between humans, and as if the technology depended only upon humans. Moreover, the practice can take place, and increasingly does, without the situational embodied presence of other humans! Phonetic reading not only emphasizes communication primarily among humans, and absent (even dead) ones at that, but also with oneself.

> The capacity to view and even to dialogue with one's own words after writing them down, or even in the process of writing them down, enables a new sense of autonomy and independence from others, and even from the sensuous surroundings that had earlier been one's constant interlocutor. The fact that one's scripted words can be returned to and pondered at any time that one chooses, regardless of when, or in what situation, they were first recorded, grants a timeless quality to this new reflective self, a sense of the relative independence of one's verbal, speaking self from the breathing body with its shifting needs. (Abram 1996, 112)

With the advent and spread of phonetic writing, human experience in the West increasingly identifies itself with the possibilities that phonetic reading suggests: the structure of experience as reflexive, the being of abstract ideas, a realm of relative security and permanence, and freedom from the complex powers of nature.

Heraclitus does not yet inhabit a literate culture. His speech, nevertheless, marks a transition from a human orientation that expands beyond humans to an awareness virtually limited to humans. One indication of this is visible in Heraclitus's description of *psyche*, in terms of *logos*.

With these ideas in mind, let us consider Heraclitus fragment D. 45: "The limits of *psyche* you will not find even if you set off on every route, so deep a *logos* does it have." In pursuit of "the limits of *psyche*," one sets "off on every route," seeking the elusive boundaries. "[S]o deep a *logos* does [*psyche*] have." Fragment D. 45 suggests that the limits of *psyche* connect in crucial ways to *logos*. *Psyche* is a *logos*. But what does this mean?—what does it mean that the term "psyche" is a *logos*? *Logos*, during the lifetime of Heraclitus,[29] has a broad host of associated meanings. If one agrees that fragment D. 45 invites one to understand *psyche* as existing a *logos*, one will want first to acknowledge the complex range of significance of "logos."

The term "logos" has at least twelve various uses in and before the fifth century B.C.E. (Guthrie 1962, 419–425). These are: "anything said or written"; "to have worth or esteem"; "deceptive talk"; "a treaty or agreement"; "a command"; "thought or reasoning as opposed to sensation"; "the truth of the matter"; "full or due measure"; "correspondence, relation, proportion"; "holding a conversation with oneself"; "cause or reason of a neuter (as opposed to personal) subject," and "general principle or rule." None of these uses is predominant, stable, and organized enough to indicate a relatively coherent concept of *logos*. Nor does any invoke a connotation that would presuppose a unified personal subject.

Keeping in view the variegated uses of the term "logos" during this preliterate-to-literate transition, it should come as little surprise that Heraclitus's *Psyche and Logos* fragments have come to mean many things. More important, in my view, is the range of meanings that the pre-Socratic term, "logos," enables in psyche contexts. That Heraclitus speaks of and to a culture, for which the term "logos," with respect to "psyche," connotes a range of meanings, is remarkable. A predominant translation of "logos" in contexts with "psyche"—in the literate Christianized West—emphasizes a certain preferred reception of "logos." Such receptions typically view "logos" as "reason, especially with respect to numbers." If "logos" in association with "psyche" had evoked a range of more disparate and sometimes equivocal significances for Homer and Heraclitus, for literate speakers, "logos" with "psyche" begins to signify an allegedly unified concept of reason (co-emerging with the unified concept of *psyche*). Gradually, the less determinate sixth and fifth century B.C.E. connotations of "logos," with respect to "psyche," would fall away.

This is especially visible in Plato's discussion of the soul in *Republic* (441e). Plato's central character, Socrates, names the three parts of the soul; he names one part λογιστικόν, translated by Paul Shorey as "rational" (Plato, trans. 1937, 407) and by John Sallis as "calculating" (1986, 369). The term λογιστικόν comes from λογιστικός, the "primary sense of which is 'computation with numbers' in a rather practical sense" (Sallis 1986, 369). Sallis shows that Plato ties together this sense of the "calculating" part of *psyche* (λογιστικόν), with the signifier "logos." "It should also be noted, however, that the root word is '*logos*,' as is especially brought out when Socrates speaks of the spirited part taking the side of the calculating part against desire: what he says precisely is that 'spirit becomes the ally of *logos*' (440b)" (Sallis, 1986, 369). By contrast, as shown above, Heraclitus fragment D. 45 ("The limits of *psyche* you will not find even if you set off on every route, so deep a *logos* does it have,") speaks of and to a people familiar with the notably more expansive and complex options for rendering "logos" in contexts with "psyche."

The overflowing and dynamic preliterate notion of *psyche* becomes all but lost with the rise of literacy and corresponding limited and unified idea of *psyche*. An example of the range of reference that would eventually cease to be triggered, when speaking or reading the term "logos" in certain contexts, is visible in Sophocles's fifth-century, B.C.E, mythical drama, *Oedipus the King* (trans. 1982). In it are at least four uses of "logos," uses which we typically do not associate with the common Latin ("ratio") and English ("reason," "calculating part") translations of the Greek "logos." One can see vividly, in *Oedipus the King*, the contexts supporting an expansive range of connotations for fifth-century B.C.E. "logos." By comparison, Heraclitus's "logos" happens in stark, cryptic, and archaeologically fragmented contexts. For the literate reader living after Socrates, and generally unfamiliar with the earlier more expansive connotations of "logos," the additional range of meanings would not come to mind, let alone seem credible.

Continuing our analysis of Heraclitus's fragment, D. 45, let us view the sixth-to-fifth century B.C.E. range of connotations for "logos" with "psyche" by examining certain preliteracy contexts from *Oedipus the King*. Like Heraclitus's speech, Sophocles' mythic tragedy, *Oedipus the King*, was spoken to a largely preliterate audience. One of the twelve meanings that "logos" can communicate is "worth or esteem." Such worth, for instance, can be said to exist in the love Oedipus shows for his parents, for whose sake he exiles himself from home; and for his children, for whom he reaches out—now a blind man—to touch and hold them, for the last time. Heraclitus's fragment, D. 45, requires us to connect the senses of "logos" to our conception of "psyche." And so, if "logos" means "worth or esteem," then according to D. 45 when "worth" is substituted for "logos," the fragment reads: "so deep a [worth] does it [*psyche*] have." In other words, one way we can interpret "psyche" within this fragment is that *psyche* has "deep worth."

Another meaning of "logos" is "speech that deceives." In association with this, the term "psyche," in fragment D. 45, would then connote "beguiling speech." We can recall, in the play, the ciphered speech of Phoebus. Phoebus *says* (*légon*) the oracle to Oedipus at Delphi. Phoebus anticipates Oedipus's tragic fate—to beget "a brood that men would shudder to behold" (Sophocles, trans. 1982, 74). The exoteric meaning of the oracle seems clear, and Oedipus reacts accordingly. From his mother and his father he exiles himself. But the esoteric significance remains hidden. Oedipus understands by "mother" and "father"—a woman and man who are not them—and by "monstrous children"—offspring with his "mother" that he did not have. Like the saying of the Delphic oracle, *psyche* beguiles.

In fragment D. 45, one can attribute to "logos," and so too, to "psyche," another meaning—the notion of "command." "Lógos" is the term Sophocles uses to refer to the tale of the oracle heard by Iocasta and recounted by the Theban (Sophocles, trans. 1982, 87). Iocasta, frightened by the Delphic vision, finds herself compelled to bring death to her child. In addition to the latter, the oracle sets forth another ordinance. It demands a fate that proves impenetrable by Oedipus—a man capable of unraveling the seemingly impassable riddle of the sphinx—and unavoidable by Iocasta: a woman who takes pronounced measures to obviate the ill fortune. The disclosing of their lives reveals to Iocasta and Oedipus a fate knit, in unfathomable ways, with the oracle's saying.

The lives of Oedipus and Iocasta show the characters bound to—even constituted by—a saying. Just as a saying can constitute one's fate, it can also constitute one's thought processes. The relationship between a saying and a person's thought processes indicates another meaning of *logos*—*logos* as "a process of reasoning" in the sense of "reasoning with oneself" or "holding a conversation with oneself." In fragment D.45, this might connote: *psyche* "so deep a reasoning with itself it has." This signification of "logos" also arises in the Oedipus myth, in which "logon" can be translated, "reason with yourself." Oedipus wrongly accuses Creon of killing the former king, Laius. In response Creon declares, "Not so, if you will reason (*logon*) with yourself, as I with you" (Sophocles, trans. 1982, 67).

My strategy of distinguishing connotations of "psyche" during Greek preliteracy, from those that come with literacy, has been guided by a particular excavation of preliterate speech generally, and examples of such in Heraclitus's fragments particularly. Pre-Socratic uses of "logos," as seen in fragment D. 45, are rich and variegated. And their complexity and connectiveness to a range of specific kinds of contexts suggests that the signification of "logos" cannot, with any ease, be identified or reductively conceptualized. By extension, such diffusion and expansiveness can be said to characterize Heraclitus's sense and use of "logos" with respect to "psyche." The term "psyche," with respect to "logos," during preliteracy seems not yet to include, let alone predominantly reduce to, a concept of a rational human self. Heraclitus's D. 45 "psyche" suggests a markedly different idea of *psyche* than that characterized in the West by a dominant reading of Plato. Heraclitus's, in contrast to "Plato's"[30] *psyche*, appears expansive, complex, and contextualized. Its signification resists the conceptual reduction—conferred through predominant Christo-Western receptions of Plato's psyche—to an identifiable and morally responsible human subject.

In *Plato's Pharmacy* (1972/1981), Derrida shows Plato's writing failing to acknowledge the inevitability of the sign, that is, the impossibility of the existence of being, apart from the sign that presents it. According to Derrida, Plato's writing[31] attempts to dominate over the inherent ambiguity of a sign like *pharmakon* (drug, remedy, poison), and to imply that for each word or sign, in a given context, there corresponds a stable referent and "correct" meaning.[32] Just as such writing opts for a logic that refuses to tolerate the equivocation of meaning for a word in a particular context, it ignores the way that being appears, that is, as a sign—and thus obscured and anything but transparent. According to Derrida, this aspect of Plato's writing privileges clear-cut oppositions: being, nonbeing; appearance, reality; good, evil—oppositions whose credibility dissolves, when understood in the presence, indeed *as* the presence, of their inevitable and opaque host: the sign. "Being" thus confounds itself as any sort of opposition to "nonbeing," Derrida suggests. And this is because the sign—the presentation—equiovocates. That is to say, it appears and exists *as* indefinite. Thus, "Being," never logically opposes "nonbeing." The same is the case for "nonbeing." The sign, therefore, defies a logic of opposition.

For some so-called Platonists and Aristotelians, philosophical writing records a supposed eternal logic and true being of the world. It inscribes a special kind of speech—speech hypothesizing timeless propositions transcending situation and context. If philosophical writing emerges at about the time of Greek literacy, it is reasonable to contend that a practice of philosophical or conceptual thinking also arises at this time. I have indicated that this is indeed the case, but not because philosophical writing enacts or discloses such an eternal logic. On the contrary, the advent of phonetic script generates the appearance of an eternal logic, and the illusion that humans are independent from a sensuous, natural world, that sustains not only our bodies and our speech, but also the appearance of the eternal logic itself.

It becomes clear that philosophy, understood as a mode of thought that presupposes the existence of abstract concepts, has not always been a human activity in the West.[33] The beginning of philosophy, then, is the beginning of a genre of writing that depends upon the emergence of phonetic writing. For this reason, one may contend that the Heraclitus fragments, which were written before philosophy takes root, ought to be interpreted by a method that avoids a primary attribute of philosophy: conceptual thinking. It is the preconceptual, preliterate character of the Heraclitus fragments that draws Nietzsche's thought toward Heraclitus's.

Heraclitus's speech may also attract Nietzsche because it registers a transition *between* preconceptual and conceptual thinking. It does so, with respect to the idea of *psyche*, for instance, by beginning to link the ideas of thinking and *psyche*. As we already saw in D. 45, "logos" connects "psyche" to cognition. This connection is significant and indeed was not made during Homeric-era speech, when "psyche" was associated with life, breath and shade, but not with attributes of a personal self, that is, with cognition, feeling, and bodily appetite.

One can see this new development in Heraclitus's fragments. "Psyche" is associated not only with cognition/speech but also with sensation and bodily affect. The *Psyche* and Bodily Senses fragments (Category II) attribute to *psyche*, certain capacities of bodily sensation. For instance, they associate with "psyche" the capacity to see and hear (D. 107), smell (D. 98), and desire (D. 85). These go beyond the Homeric association of "psyche" with life or breath, by including the human bodily senses. They also, like the *Psyche*, Dry and Wet fragments (Category III) suggest that *psyche* can fluctuate between better and worse states. The soul is "most efficient and best" when it is dry (D. 118). And when one must "fight against desire" and loses, *psyche* pays for it (D. 85).

By excavating the expanse of the signification of "psyche" in the speech of Heraclitus, one sees that the notion of *psyche* appears in relation not only to cognition (D. 45) but also to bodily sensation. This is not to say that the term "psyche" of Heraclitean speech refers to a concept of a human subject defined by the unity of such processes. It indicates instead a relation—still dynamic—between the ideas of *psyche*, cognition, and other *logos* connotations. Although "psyche" here does not yet appear as a supposed stable concept, it has shifted, beyond the Homeric connotations of life and death, to affiliate with thought, desire, and sensation, attributes brought together in Plato and Aristotle's concepts of *psyche*.

Eric Havelock's work shows that speech, during the Homeric era of preliterate Greece, differs in important ways from that spoken or written during the Classical era of early literate Greece. Some attributes of preliterate speech include vivid appeals to sensory imagery, infectious rhyme-schemes, and organic story structures.[34] Another attribute, one that I examined in the speech of Homer and Heraclitus, is a preconceptual use of the term *psyche*. Most research indicates that the earliest uses of "psyche" that intend a unified concept of human soul, occur in the late fifth century B.C.E., that is, they coincide with the lives of Socrates and Plato. Indeed, my examination of "psyche" in the speech of Heraclitus shows a transitional use of the term. Although this use of "psyche" remains generally preconceptual, it nonetheless is beginning to be associated with cognition, feeling, and certain bodily organs—attributes that will participate in a connotation of "psyche" in the writings of Plato and Aristotle, and which is predominantly received by Socrato-Christo-Western interpreters as a conceptual unity.

"Psyche" is a term that comes to signify something like "human nature." By focusing on the genealogy of the term "psyche," from a preconceptual meaning to an apparently conceptual one, I am able to show how speech transforms with the rise of literacy. More precisely, through a genealogy of our perceptual structures, I show how speech eventually gives to perception the appearance of identifiable concepts.

The perception of such concepts correlates with the view that human experience and so-called 'human nature' is independent of the very life world that seems to sustain it: the earthly and the mineral; the photosynthesizing and the creaturely (Abram 1996, 93–136; 225–260). Just as the rise of phonetic writing structures human speech and perception in ways that appear to no longer depend upon contextual details of terrain, vegetation, climate, and animal life, so too, it would seem, it encourages one's perception of selfhood to duplicate such independence. Perhaps it should not be a surprise that an apparently coherent concept of *psyche*—an idea about substances as theoretically and actually distinct from the details of their environment—would emerge with literacy. This chapter is able to make this case with respect to the development of the abstract concept of *psyche* or self, by examining connotations of "psyche," which co-arise with literacy. Although this project hones in on the emergence of the abstract concept of *psyche,* during the era literacy is established, it also suggests the contemporaneous arrival of conceptual thinking generally.

Heraclitus's speech, which was spoken *before* the human technology of phonetic script takes root, ought to be interpreted by a method that avoids certain assumptions of phonetic writing in the West. This chapter has exposed three assumptions of this kind. These are 1) that human speech and perception naturally, or correctly, operate independently of a more-than-human world; 2) that human speech and perception are naturally and necessarily experienced through a subject-object grid; and 3) that human perception of a unified concept of human *psyche* or self—a self independent from a more-than-human world—exhausts pertinent experiences of self-understanding. These assumptions are best set aside if one is to make out the apparent limits of—and begin to see the expanse beyond—human technological thinking.

The speech of Heraclitus is particularly helpful in registering not only the possibility but also the actuality of speech and perceptual patterns not yet dominated by the three assumptions above. It is the possibility and actuality of such experience that could be said to attract Nietzsche to Heraclitus's speech. Heraclitus's speech shows to us and to our own habits of literate speech and perception, ways of being that are beyond our current deciding limits. By preserving structures of an earlier preliterate speech, Heraclitus's speech registers in human speaking and perceiving a more-than-human world. It also foreshadows a human speech to come, one that virtually obliterates from human awareness its persisting dependence upon the sensory particulars of its surroundings.

PART VI. LIVING AND DYING SPEECH

From the shores of our own lands and cultural literacies, Nietzsche gazes towards the sea and the faraway place of Heraclitus. Nietzsche turns towards speech beyond literacy— pointing himself and us toward a promise. If Nietzsche points us in a direction, he gives, at best, a cryptic map of how and why we ought set out. In *Twilight of the Idols* (1968b), Nietzsche's first two aphorisms, replete with life and death imagery, link philosophers especially with the latter: imagery of death. Heraclitus is an exception. Perhaps Nietzsche's Heraclitus is best associated not only with the imagery of life, but also with another kind of imagery: the imagery of war. In the preface Nietzsche writes: "War has always been the great cleverness of all spirits who have become too inward or too deep; in the wound itself yet lies healing power" (KSA 6, 57). War here appears as a great reviver of spirits. Spirits that have penetrated too far "inward," too "deep" into themselves, need to find a way beyond themselves, or at least to less shallow ground. War wounds; and wounds shock, impair, and sometimes traumatize. They also stimulate circumstances for transformation and renewal. Nietzsche's *Twilight of the Idols* could be said to speak a fight, for life, and for renewal.

Its images of strife and wounds send a message about war not often heard. War maims but it also renews. Indeed Heraclitus, Nietzsche's perceived comrade in war, offers a number of fragments similarly touting the revitalizing power of conflict and war.

D. 53 War is father of all and king of all. (K #83)
D. 48 The name of the bow [*βίος*] is life; its work is death. (K #79)
D. 80 One must realize that war is shared and conflict is justice, and that all things come to pass (and are ordained?) in accordance with conflict. (K #82)
D.125 Even the potion separates unless it is stirred. (K #77)
D.11 All beasts are driven by blows. (K #76)

The preface to *Twilight of the Idols* gives us an image of an injured spirit who has turned too far into her- or himself to live well. It is not difficult to see who, for Nietzsche, fits this description in *Twilight*'s early aphorisms: the philosopher, especially Nietzsche's version of Socrates, who is a sick man—dissector of instinct and soul.

Would Nietzsche wish to genuinely see himself exempt from this illness of inwardness? The *logos* and perceptual structures his literate, philosopher ancestors pass down to him, register a human technology of increasing constriction. Its boundaries are circling an especially human world. The signifier "logos," today, in an era of protracted enlightenment values, is 'efficiently' translated, "rationality." The many sixth-century B.C.E. associations and dependencies of *logos* seem for most of our contemporaries relics of a generally dead speech.

When we, of a technological West, are asked to distinguish a human *psyche*, or subject, from other beings and creatures, commonly we attribute to our selves, "rationality." The Greek word "logos," in the context of "psyche," in Aristotle or Plato's texts, has dominantly been understood as "rationality," too. In the light of the transition to literacy, we can see forces that would push interpreters of Plato and Aristotle to understand their respective notions of *logos* and *psyche* this way. Nietzsche seems to have noticed this long-standing turning-back-upon-itself of human speech in the West. Human speech draws inward, recursively circling in itself that which becomes, increasingly, merely itself, or what Abram calls a merely human world. Human technological reason tends to fasten to its own reflection, to focus its knowledge more narrowly, and as Nietzsche sees it, to deepen illness.

Nietzsche's *Twilight of the Idols* reflects such entrenchment. Its early aphorisms allude to a Socrates who stultifies himself with a his own belief in human reason, an identifiable soul, and "concept-mummies" (TI, RP 1). The form of Socrates's speech and perceptual structures resemble parts of *Twilight*'s form—its thick aphorisms replete sentence after sentence with strikes pounding away at philosophers, but only as a *philosopher* can do: with sporadic retreats into sustained reason—into arguments. *Twilight*'s blithe sayings of "Maxims and Arrows," on the other hand, relate to the aphorism like seas off the shore of a literacy-bound land. Reminiscent of Heraclitus's cryptic fragments, they strike, stimulate, and slip, leaving one with a plurality of interpretive options rather than a more coherently discernible message. Like Heraclitus's fragments, these too are curt and provocative; they press an opening of technological thinking. Like seas arriving from distant lands, they beckon to us, disclosing—as if to herald—forgotten opportunities for tired speech.

It is *against* such seas, that Nietzsche sees Socrates raise his philosopher's stick. Nietzsche's Socrates, in *Twilight of the Idols*, personifies diminishing life and speech. This version of Nietzsche's Socrates pledges allegiance to theories about being: its permanence and transcendent truth. Such commitments Nietzsche interprets as the last resort of an otherwise defenseless, life-threatened individual. "Dialectics can be only a *last-ditch weapon* in the hands of those who have no other weapon left. One must have to *enforce* one's rights: otherwise one makes no use of it. That is why the Jews were dialecticians . . . and Socrates was a dialectician too?" (TI, "The Problem of Socrates"[35] 6). Nietzsche's Socrates here, seems at war. In his desperation, suggests Nietzsche, Socrates armors himself too heavily. No blows can wound his wrapped discourse. As Socrates fends off vulnerability to new attack, he also cuts off access to a wound sinking inward that he lives. Is not Socrates, asks Nietzsche, "an expression of revolt? of the *ressentiment* of the rabble?" (TI, PS 7).

Nietzsche's own speech is a descendent of that which he attributes to Socrates. And so, he suggests, he too suffers from captivity. His speech, too, is tight in the wraps of technological self-understanding. Unlike Socrates, though, Nietzsche sees the inward-turning wound he harbors. It is this wound that Nietzsche's text not only dresses but redresses. In so doing, it aerates a field that had been sealed off by inherited Socratisms. It opens Nietzsche's self-reflexive speech to the speech of Heraclitus—to the off-shore air, sunlight, and secret seas—for renewal and recreation.

Nietzsche's nod toward Heraclitus's speech is an affect, which in effect puts into question the unexamined assumptions structuring the speech and perception of literate culture. My analysis of human speech and perceptual structures existing prior to the spread of phonetic writing, suggests that the structures shaping speech since literacy are not the only ones pertinent or available, even if our cultural habituation makes difficult accessing others. Acknowledging the possible, if difficult, entry to other perceptual structures, can open worlds. Nonetheless, many today presume that contemporary human speech and perception reflect, plainly and simply, human progress—benefits with few costs. Widespread is the view that our practices and ways of being have advanced beyond the experiences of our preliterate ancestors. Whether or not true, the idea that the experience of preliteracy would have little to offer human beings of the third millennium, I would contend, is not. One common way many of us display our belief in human progress, is by distinguishing ourselves from other species and beings. We are unique, this traditional view suggests, because we think abstractly. And, we do think abstractly! But we remain, nevertheless, generally unaware that thinking abstractly is made possible only once we have nurtured the *appearance* of independence from nature—which is to say, turned away from nature. Our pervasive faith in progress, and in ourselves, has the form of a tautology. We reorient ourselves, toward only ourselves (rise of literacy), and afterward we declare that we are, quite naturally, dependent only on ourselves (legacy of literacy). Structured into our lived practices of such apparent independence, is a subject-object sentence structure. And its structure, in turn, provides for the perception of the discreteness of beings, especially human beings.

Such subject-object reflexivity softens the soil; and a predominant Christo-Socratic perception of self takes root. Through it, many perceive humans to be hierarchically superior to, and discrete from, other living creatures. Eventually, many of us would come to see humans alone as the creature that persists even after the body dies—or rather that one's *psyche* or soul does. The capacity in literate speech for conceptual abstraction enables us to see ourselves as separable, at least in theory, from a more-than-human world. Indeed, since the spread of literacy, it is precisely such a capacity for rational abstraction that has come to dominantly define human subjectivity. If this remains the predominating definition of subjectivity in the West, such a definition of self is certainly not Nietzsche's.

Nietzsche's gesture to Heraclitean speech initiates in Nietzsche's own, an opening. The opening reaches toward realms that confound and expand pervading conceptions of human self-understanding. Speech since literacy can be said to carry in itself, if also to cover over, traces of its own preliterate heritage. Nietzsche's writing mobilizes structures that are implied by the ancestry of his and our speech. Interestingly, this mobilizing in and of our perceptual structures can be said to move not like an agent upon passive structures absolutely outside itself, nor like a subject collecting forces strictly outside itself to stimulate change within. The preliteracy structures, if muted, still remain in the ground of his and our own speech and ways of existing. Nietzsche's mobilizing speech can be said to move according to a structure exceeding inside and outside and subject and object. It transpires much like a mode of the middle voice.

Notes

1. INTRODUCTION: NIETZSCHE AND EMBODIMENT

1. See Bernd Magnus, Stanley Stewart, and Jean-Pierre Mileur's *Nietzsche: Life as/and Literature* (1993) and Alexander Nehamas's *Nietzsche: Life as Literature* (1985).

2. Herman Siemens describes the importance of the notion of fiction for Nietzsche in terms of the "open horizon" it provides and its "performative aspect." "The distance between teleology and fiction is measured by the difference between enclosing the horizon of the future, and playing with an open horizon" (Siemens 2001, 81). What Siemens calls "agonal transvaluation"—a dynamic of "'saying and unsaying' (Blondel)," empowering and disempowering—is fostered by a strategy of fiction's performativity: "Nietzsche's philosophical discourse at the surface of the text is but part of a larger organization or economy of energy, grounded in embodied, affective engagement" (2001, 82).

3. See n. 6, chapter 2.

4. Jean-Jacques Rousseau's *Emile: or On Education* (1762/1979, 361) says that nature positions woman as child bearer and caretaker.

5. Moira Gatens's concept of body in *Imaginary Bodies: Ethics, Power, Corporeality* suggests such a contradiction.

6. Here, with some hesitation, I compare the Buddhist idea of "emptiness." Nietzsche's description of a modern experience and concept of "nihilism"—of no proper self, of no God, of suffering the apparent meaninglessness of suffering—is like that of "Great Doubt" (Nishitani 1982, 18–19). Paralleling the transition in certain Buddhist traditions from "Great Doubt" to the experience of "absolute emptiness" is the Nietzschean transition from nihilism to *Ja-sagen* [Yes-saying]

(Nishitani 1982, 66, 124). "On the field of emptiness, all things appear again as substances . . . though of course not in the same sense that each possessed on the field of reason" (124). Because having passed through the field of nihility, the "selfness of the thing as it is on its own home-ground . . . cannot be expressed simply in terms of its 'being one thing or another'" (124). That is to say, it cannot be expressed according to ordinary language of reason. This concept and experience of absolute emptiness Nishitani also calls the "Great Affirmation, where we can say Yes to all things" (124).

How far to take this parallel remains a question for another project. For now let us note that Sumio Takeda (2000–2001, 105) and Keiji Nishitani (1982, 66, 124) conclude that the concept of emptiness surpasses that of Nietzsche's *Jasagen* both with respect to the absoluteness of its nihilism and its affirmation of being. Neither, however, considers Nietzsche's view of the human tendency to self-deceive, which would make Nietzsche wary of any absolute ontology. This, together with Nietzsche's view of "great health" not as a settled ontology or possession but "a struggle, a process" (Glenn 2001, 110), that one "'acquires continually, and must acquire because one gives it up again and again, and must give it up'" (GS 382, qtd. in Glenn 2001, 110), leave open for further inquiry the relation between Nietzsche's thought and that of certain Buddhist traditions. For further discussion see Steven Laycock (1994) and Carl Olson (2000).

7. Judith Simmer-Brown (2001a).

8. The language "as metaphor and through metaphor" comes from Eric Blondel (1991, 205).

9. See, for instance, Tim Murphy (2001).

10. By phenomenological approach I mean variations of a method first described by Husserl and generally involving a suspension of one's judgment about correct ways to characterize being or beings. One substitutes for judgment observations about the way being or beings appear for human being.

11. See n. 4, chapter 7.

12. See n. 5, chapter 7.

13. See n. 2, chapter 7.

14. Merleau-Ponty's primary resource for case studies was research by neurologist Kurt Goldstein, who examined and wrote extensively about the behavior of men injured in World War I. Goldstein's major work, *The Organism*, was first published in German in 1934 and appeared in English in 1939.

15. The term "make up" is from Elaine Scarry (1985).

16. Nietzsche's characterization of "Socratism" arises in his writings as early as *The Birth of Tragedy* (1967).

17. I borrow this phrase from David Abram's *The Spell of the Sensuous: Perception and Language in a More-Than-Human World* (1996).

18. See Scott (1990, 19).

19. See Robb (1986, 317).

20. Fiona Hughes attributes to Nietzsche the necessity "of a particular life" of an individual. "In experimenting with life, a new life is created" (Hughes 2002, 131–132). Individuals have the freedom to be "experimenters," to "accept neither the jurisdiction of external laws nor the necessity of an alien nature (131). "[T]he recognition of a necessity will reveal the complex dynamic—interestingly similar to negative dialectic—that is the real structure of reality" (131).

2. OPENING NIETZSCHE'S GENEALOGY TO "FEMININE" BODY

1. By *self* (or subject) I mean human self-understanding. I will trace Nietzsche's view of the transformation of the self/subject from a disunified subject that Nietzsche calls the "legal subject" to a supposedly unified subject. A unified subject has intelligent desire and contrasts with the disunified subject whose desire lacks the feasibility of such intelligence. Judith Butler explains this: "If desire were a principle of irrationality, then an integrated philosophical life would be chimerical, for desire would always oppose this life, undermine its unity" (Butler 1987, 3). I write "supposedly unified subject" because Nietzsche's tendency is to reveal the infeasibility of intelligent desire. For Nietzsche, the object of human desire, the "Good" is not, as for example, Kant, universally recognizable according to rational law.

2. Nietzsche attempts to offer a non-dualist conception of mind-body relation. Eric Blondel gives a compelling view of this in *Nietzsche: The Body and Culture* (1991). For Blondel, Nietzsche's non-dualism points to body as a non-foundational primacy. "Spirit and body constitute one unit, but it is a plural unity" (Blondel 1991, 206). "The body is therefore an intermediary space between the absolute plural of the world's chaos and the absolute simplification of intellect. If chaos and humanity begin with the body as interpretation, a model of this unity and plurality, a schema of the will to power, if the body comes first, it does so as a mode of the *mixed*. If first (as multiplicity) *in relation to the intellect* conceived of as unifying and simplifying, while it will come second *in relation to the chaos of the world*. Nietzsche, therefore, intends not to reduce the intellect to the body, but, in presenting the *body* as a 'plurality of intellects', to reveal the radical nature of *plurality*" (207).

3. See n. 1.

4. I translate *Rechtssubjekt* as "legal subject" (KSA 3, 298).

5. My use of "unified subject" corresponds to Nietzsche's *reaktive Mensch* or *Mensch des Ressentiment* that metamorphoses out of the earlier *Rechtssubjekte* (legal subject) in the "Second Essay" (KSA 5, 311). See n. 1 for definition of "unified subject."

6. I have chosen the term *constitution* instead of *item*, *object*, or *thing* to point to the changing character of things, or rather, to their resistance of reification and metaphysical identity. Indeed, what I characterize below as Nietzsche's *relation*, shows Nietzsche's view of the formation of semantic signification and subjectivity. This *relational* process of formation undermines a notion of reified things, which presupposes metaphysical identity.

7. Compare to n.2.

8. Gatens can introduce such dynamism into the otherwise seemingly static categories "sex" and "gender" through a Spinozist monism. As Gatens notes, "both the extensive (bodies) and the intensive (minds) are conceived by Spinoza as complex fields of interconnecting powers and affects" (149).

9. See Oliver's "Opening and Closing the Possibility of a Feminine Other" for why Nietzsche's discourse of body is "always of a masculine body" (1995, 16–25).

10. Nietzsche, not surprisingly, testifies to the traditional association between women and cooking. He not only links women to cooking, but "woman"—and with quite the sharp tongue. "Stupidity in the kitchen; woman as cook: the gruesome thoughtlessness to which the feeking of the family and of the master of the house is abandoned. Woman does not understand what food means and wants to be cook" (BGE 234).

11. I understand Foucault's notion of *discourse* to be similar to what I call Nietzsche's *relation*. Foucault dispatches *discourse* more explicitly than Nietzsche exposes *relation*. Moreover, Foucault's *discourse* is probably a descendant of Nietzsche's *relation* (Foucault 1971/1977, 139–164).

12. Carol Diethe translates *Verwandtes* as "Related Matters" (1994, 38).

13. Nehamas helps collect many of the relevant passages (1985, 100–105).

14. Edward Andrew points out in his genealogy of the language of "values" that Nietzsche deliberately exports value-discourse from economic-discourse. Prior to Nietzsche, from "Plato to Mill, the language of values was confined to economics" (Andrew 1999, 66). Andrew criticizes Nietzsche's wholesale use of the language of values in discussions about virtue, aesthetics, happiness, and justice because it conflates market exchange-values of "material needs" with moral-values of "conscious choices" (1999, 73). This, writes Andrew, weds Nietzsche's values-discourse to the kind of capitalist mediocrity from which Nietzsche wants to move away (65), and prevents Nietzsche from committing to the "priceless dignity of human beings" (72).

15. Some might view guilt as a primarily psychological phenomenon. Because of its relative *Schulden* (a socio-physical constitution) and the history of physical cruelty that Nietzsche suggests contributes to the formation of *Schulden*, and so too to *Schuld*, I categorize guilt as a psychosomatic constitution.

16. I modify Kaufmann's translation. Kaufmann translates *das versprechen darf* to "with the right to make promises."

17. Though Nietzsche implies that human preference, taste, and power relations have the power to affect and change the possible interpretations and nature of human being, Nietzsche does not suggest that humans are free of necessary laws. In terms of human nature, we are "unfree" if by "free" one means "the foolish demand to change one's *essentia* arbitrarily, like a garment and "free" if by "unfree" one means one's nature is bound by an unchanging essence (PTG 7).

18. *Heranzüchten* is a compound word combining *heran* and *züchten*. Like many compound German words, it is not to be found in most dictionaries. In the case of *heranzüchten*, it has virtually the same meaning as its root word: *züchten*. *Heran* is an adverb that means "to come on or along"; or "to come close or near to something." *Heranarbeiten* means "to work one's way along." *Züchten* means "to breed" in the sense of harvesting pearls or farming a crop or animals. Thus, *heranzüchten* derives most if not all of its meaning from *züchten* because *züchten*, as a process of formation, also implies a gradual coming along. For this reason, I explore the etymology of *züchten* without concern for *heran*.

19. Nietzsche's value-discourse, writes Andrew, requires Nietzsche to disavow the "priceless dignity of human beings." To assume the "priceless dignity of human beings," Andrew must ignore human experience per se. This is because, as I show in chapters 2 and 3, human experience per se (as distinguished from conceptual assumptions about experience) shows that (moral) valuations about objects of experience are co-configured with and by the objects of experience.

20. For a definition of *feminine symbolic* and explanation of why I choose curry as a metaphor for a symbolically feminine body, see my introductory comments.

21. Because the term *curry* is Western in origin, its signification does not translate into the tradition of Indian cuisine except in the case of those of Indian descent who have been influenced by the Western concept.

22. Double genitive refers to the double meaning of any genitive phrase. In genitive phrases, the grammatical agent and patient are ambiguous. For instance, in the genitive phrase "abuse of politics," politics can be the agent of abuse and abuse the object or vice versa. If citizens use intimidation to keep certain voters away from the polls on election day, the political electoral process could be said to be the object of abuse. At the same time, the political strategy of intimidation illustrates a politics that abuses certain citizens. From this angle, politics is the agent that abuses.

23. To the human whose conscience does not clear itself, Nietzsche attributes *ressentiment*. Ruth Abbey offers an interesting genealogy of Nietzsche's concept of *ressentiment*. She argues that Nietzsche's use of the concept of vanity (*die*

Eitelkeit) operates not as a precondition for the superhuman, as had been argued (Donnellan 1982), but as precursor to Nietzsche's idea of *ressentiment*. Both *ressentiment* and vanity she argues are linked with a dearth of self-love (Abbey 1999, 59).

24. KSA 5, 298.

25. Forgetful mind names embryonic conscience. Conscience emerges (after much cruelty) with the advent of memory.

26. Implicit in Nietzsche's *relation* is the impossibility of strict definition. This arises in the loss of strictly identifiable "things."

27. The second reason is Nietzsche's hypothesis of the "blonde beast." The one whose "instincts of wild, free, prowling man" threatened the organizing tactics of the "blonde beast" was "incarcerated within and finally able to discharge and vent itself only on itself: that and that alone, is what the *bad conscience* is in its beginnings" (GM 2, 17).

28. The phrase *relation without relata* refers to relationality in the absence of absolutely definable items to relate (Magnus & Higgins 1996, 7). I agree with Bernd Magnus, Stanley Stewart, and Jean-Pierre Mileur, that many Nietzschean concepts are self-consuming and "relation without relata" not only names this logic but also exemplifies it. The ground that constitutes such a concept, points to, at the same time, a disintegration of that very ground. Such a logic of self-destruction is typical of Nietzsche's thought, which attempts to think through and beyond the tradition and logic of Western philosophy (Magnus, Stewart, & Mileur 1993, 21–34).

29. Nietzsche periodically speaks of the stomach, intestines and digestion. One example: "[The spirit's] intent in all this is to incorporate new 'experiences,' to file new things in old files— . . . all of which is necessary in proportion to a spirit's power to appropriate, its 'digestive capacity,' to speak metaphorically—and actually 'the spirit is relatively most similar to a stomach" (BGE 231).

30. See n. 17.

31. Rousseau's belief in proper "natural" man and woman compares to his view of the proper meaning of words. Jacques Derrida discusses this in "The Violence of the Letter: From Levi-Strauss to Rousseau" in *Of Grammatology* (1967/1974). "Rousseau no doubt believed in the figurative initiation of language, but he believed no less, as we shall see, in a progress toward literal (proper) meaning. 'Figurative language was the first to be born,' he says, only to add, 'proper meaning was discovered last' (*Essay [on the Origin of Languages]*). It is to this eschatology of the *proper* (*prope, proprius*, self-proximity, self-presence, property, own-ness) that we ask the question of the *graphein*" (107).

32. Gatens's concept of body in *Imaginary Bodies* (1996) suggests such a contradiction.

3. NIETZCHE'S ASCETIC IDEALS AND A PROCESS OF THE PRODUCTION OF EMBODIED MEANING

1. The literature has many accounts of the theme in Nietzsche's work of an apparent absence of metaphysical presence and impossibility of a logic of identity. For a discussion about fractured identity regarding concepts and proper nouns see Michel Haar (1979).

Sarah Kofman (1988) speaks of non-identity in terms of Baubô, a figure of the female deity: "The figure of Baubô indicates that a simple logic could never understand that life is neither depth nor surface, that behind the veil, there is another veil, behind a layer of pain, another layer" (197).

Gary Shapiro (1990) writes of a *translation* of the theme of non-identity from *On the Genealogy of Morals* in Foucault's *Discipline and Punish* (1975/1979); *History of Sexuality* (1976/1980; 1984/1990; 1984/1988) and Derrida's *Of Grammatology* (1967/1974) in his article "Translating, Repeating, Naming: Foucault, Derrida, and *The Genealogy of Morals*."

For consideration of the theme of metaphysical identity with respect to Nietzsche's idea of will to power see Charlene Haddock Seigfried, "Why are Some Interpretations Better than Others" (1975); Alphonso Lingis, "The Will to Power" (1979); Michel Haar (1979, 8–12); Alexander Nehamas, *Nietzsche: Life as Literature* (1985); Bernd Magnus, "Nietzsche's Philosophy in 1888: *The Will to Power* and the *Übermensch*," (1986) and James A. Leigh, "Deleuze, Nietzsche and the Eternal Return" (1978, 213–216).

For a discussion of non-identity as the figure of Dionysus in excess of limits, see John Sallis, "Dionysus: Resounding Excess" (1991). To consider Nietzsche's writing as a "smooth harmony of disparate or dissonant themes" and an "unsaying of what it says" see Babette E. Babich, "Nietzsche and the Condition of Postmodern Thought: A Post-Nietzschean Postmodernism" (1990b). For a reading of the theme of non-identity with respect to the *Übermensch*, see Bernd Magnus (1986). For Nietzsche rejecting the possibility of binary oppositions, see Alan Schrift, "Nietzsche and the Critique of Oppositional Thinking" (1989).

Some articles about Nietzsche and the (plural) production of meaning include Karsten Harries, "The Philosopher at Sea" (1988); Dick White, "Heidegger on Nietzsche: The Question of Value" (1988); Alan Schrift, "Genealogy and/as Deconstruction" (1988) and Jean-Michel Rey, "Commentary" (1988).

2. Plurality here is an homonymous and not an analogical multiplicity.

3. Kelly Oliver makes a similar point in Chapter One of *Womanizing Nietzsche: Philosophy's Relation to the "Feminine"* 1995, 17–18; in an interesting argument, Christopher Janaway says that the "Third Essay" is an exegesis not—as has often been argued—of its preceding epigraph about woman wisdom loving a warrior, but of the essay's first aphorism (1997, 254).

4. See chapter 2, n. 6.

5. Although feminist philosophers in the Anglo-American tradition have also avoided including a concept of biology in their definition of body, here I mean especially feminist philosophers working in the French psychoanalytic tradition. I include in this lineage Luce Irigaray, Chantel Chawaf, Julia Kristeva, Helene Cixous, Moira Gatens, Judith Butler, Monique Wittig, and Kelly Oliver.

6. The language of "values" used in contexts of morals or aesthetics has been shown to be a unique development of Nietzsche. According to Edward Andrew (1999), "from Plato to Mill, the language of values was confined to economics, and did not compete with, let alone take over, languages of virtue, or right, or principle or of happiness" (66). For a genealogy of the language of values, see Andrew's "The Cost of Nietzschean Values" (1999).

7. Moral valuation and definitive valuation are two subcategories of ideational constitution.

8. Cf., *Metaphysics* 1031a11–14 (Aristotle, trans. 1979).

9. Cf., Sarah Ruden, "Staring at Yellow and Green (Life and Art and PMS)" 1998, 409.

10. I find it interesting that in Edward Andrew's criticism of Nietzsche's language of values, a critique that implicitly requires a determinate dualism between subject and object, that Andrew's language suggests the opposite: "Above all we are not aware of how we are used by the language of values when we think we are using it (1999, 70).

4. NIETZSCHE'S ASCETIC IDEALS AS A PROCESS OF THE PRODUCTION OF MEANING

1. Cf., Wolfgang Müller-Lauter's reading of the will to power as both one and many. "So ist der Wille zur Macht zugleich Eins und Vieles . . . Wir müssen die Einheit des Willens zur Macht als organisierte Vielheit von Willen zur Macht verstehen" (1974, 79).

2. Babette E. Babich notes the various descriptions Nietzsche gives the ascetic ideal and the numerous oppositional positions it constitutes at different points in the text. Via such techniques, writes Babich, Nietzsche "directs a range of possible readings for the reader" (1990a, 71). Nietzsche's "truth," says Babich, in all its multiplicity, deception and disunity is one that Nietzsche's readers also seek. Including references to Heidegger's *Nietzsche* vol. 1 (1961, 28), she describes the polyvocity of Nietzsche's "truth" as akin to Heidegger's idea of "originary 'truth' . . . Like a woman—in Nietzsche's words or like a lover . . . the truth of nature as becoming, as chaos. . . . What is more, as Heidegger tells us, because what is revealed in truth is ever only partially laid open, it remains eternally

obscure" (1990a, 73). Therefore, says Babich, aligning herself with Heidegger, that which is questioned in thought must "ever be held open: the question recalled" (1990a, 73). The ceaseless seduction of questioning Babich calls "true desire." (1990, 71–73).

3. "In it [the ascetic ideal] suffering was *interpreted*; the tremendous void seemed to have been filled" (GM 3, 28).

4. The term *mean* registers doubly as "a means to" the end; and "signifying" the end.

5. Cf., Nietzsche's description in the "Third Essay" (3, 6) of the different effects of "the beautiful" on the will for Schopenhauer as opposed to for Stendahl.

6. Nietzsche quotes Arthur Schopenhauer. This corresponds, in the E. F. J. Payne translation, with the "Third Book," §38:196.

7. 1) It acts as an expedient or means to the end; 2) it signifies the end, that is, the particular end is its referent.

8. One could make parallel arguments about the genealogist's dominantly negative valuation of the ascetic ideal in relation to Schopenhauer and the ascetic priest's respective positive valuations.

9. In the "First Essay" of *On the Genealogy of Morals*, Nietzsche distinguishes a master as one who creates and defines the value of "good" out of one's own overabundence of energy, strength, and confidence and "bad" in opposition to this (GM 1, 2).

10. Garth Gillan suggests a relatedness among moral valuation, body, and power: "The belief in the body delimits the terrain where conflict can install itself, where the 'For and Against'of values can assume concrete figures in the opposition of classes, races, values and moralities" (1988, 135).

11. Leiter agrees that Nietzsche's own metaethical position precludes any epistemologically privileged view of values. This leads him to an interesting interpretation of why Nietzsche says he will be understood by few: "Now if we assume that Nietzsche does not believe his own evaluative perspective enjoys any privilege over the morality he revalues, then it would, indeed, make sense for Nietzsche to want to circumscribe his audience to those who share his evaluative taste: to those for whom no justification would be required, those who are simply 'made for it,' 'those whose ears are related to ours,' who are 'predisposed and predestined' for Nietzsche's insights. For such an audience, one's values require no epistemic privilege to nonetheless carry the weight they must carry for Nietzsche's critique of morality" (Leiter 2000, 11).

12. One who traces the development of a Western value system (slave morality) out of a different moral system: master morality and its lineage.

13. An *ideal* implies a relation between a value (good) and an entity, behavior, or activity (a chair, humility, serving others).

14. Keith Ansell-Pearson discusses the technology of cyberspace as a prevailing contemporary form of the ascetic ideal. Its "cult" status indicates a new form of the people's faith (1997, 2).

15. Compare with Charles Scott's trace of a conflating subject and object in an antiquated verb structure of early Sanskrit and Ancient Greek called "the middle voice" (1990, 18–35).

16. Traveling no more than 1,225 yards on the Sabbath was the rabbinical prescription of a "Sabbath's-day distance." Onions, C. T. with William Little, H. W. Fowler, J. Coulson. *The Shorter Oxford English Dictionary on Historical Principles*, 3d ed., Oxford: Clarendon Press, 1973, p. 1869.

5. NIETZSCHE ON A PRACTICE AND CONCEPT OF GUILT

1. This tension occurs between unity (God, morally integrated subject) and disunity (conscience that forgets, legal subject).

2. See n. 1 and n. 4 chapter 2 for a description of "disunified legal subject."

3. See n. 1 and n. 5 chapter 2 for a description of "unified subject."

4. In the "Second Essay," Nietzsche uses quotations with the term *bad conscience*, so that it typically reads *"bad conscience."* The quotations could be said to call attention to the oddity of the term and suggest a "so-called" or "if you will" in conjunction with it.

5. One will note that "bad" is the pejorative term used by the master moral system and "evil" by the slave moral system. The genealogist, generally favoring the earlier value system, appeals to it in making his moral valuation about conscience once humans internalize guilt.

6. Siemens (2001) does not deny that Nietzsche repeats the behavior and logic of *ressentiment*, but rather, "that in repeating these motions, Nietzsche remains locked in a reactive mode of evaluation" (76). He offers an intriguing rendering of "agonal transference" as therapy by comparing Nietzsche's concept of transformation (*Übertragung*) with Freud's concept of transference. "As such, transference allows for the compulsive repetition of the repressed in the distorted form of neurotic symptoms. . . . Precisely because of the freedom it gives to repetition compulsion, transference is a privileged site for the manifestation of repressed or "forgotten" pathogenic impulses and their transformation through the work of remembering" (88).

6. NIETZSCHE, METAPHOR, AND BODY

1. Some have argued that Nietzsche's thought makes no such transition but rather, has a consistent vision throughout. A recent version of this view is offered by Christoph Cox (1998). The problem with Cox's argument rests in the

selectivity he need exercise with Nietzsche's early writings. Cox focuses on *Philosophy in the Tragic Age of the Greeks*, a text that Nietzsche chose not to publish and ignores *The Birth of Tragedy*, a published text and laden with evidence of a Kantian-Schopenhauerian influenced metaphysics. I agree with such scholars as Ernst Behler (1995, 3–26) and Philippe Lacoue-Labarthe (1971), who see Nietzsche's "On Truth and Lying" marking a clear transition from his metaphysics and theory of language in *The Birth of Tragedy*.

2. Bodily feelings and affects as representations of external world forces can be said to be metaphors for that which they translate. They are "metaphorical" in the Aristotelian sense (Aristotle, trans. 1941, 1457b), which distinguishes the concept from the metaphor. For Aristotle, metaphors unlike concepts, figuratively (not properly or logically) correspond to that which they represent.

3. The language of "saying and unsaying" comes from Eric Blondel (1991, 30–33, 36, 248, 258).

4. Whether Nietzsche intends to avoid a materialist reduction even as late as 1887 and 1888 remains undecided. For a careful analysis discussing how Nietzsche's writings increasingly tend toward a reductive biologism see Gregory Moore's *Nietzsche, Biology, Metaphor* (2002).

5. Shannon Sullivan's concept of "transactional bodies" is a helpful nonreductive, non-dualist description of experience. "While nature and culture, bodies and environments, are not capable of being what they are apart from one another, this does not mean that they are identical to one another. To understand bodies as transactional is to understand them as something different from culture, even as they are culturally configured and even as they in turn configure cultures and environments. Thus, one can speak fruitfully of humans as distinct from trees, rocks, and rivers even as one recognizes the constitutive impact that each has on the other" (Sullivan 2001, 3).

6. "On Music and Words" comes from Nietzsche's unpublished fragments writeen circa 1871.

7. This analysis of Nietzsche's early writings was initially inspired by Sarah Kofman's essay "Metaphor, Symbol, Metamorphosis" (1993).

8. Examples of this more secondary sense of body for Nietzsche would be the experience of one's body as an object; the position (as opposed to tone) of one's organs when sounding out a consonant; a written symbol or word and an abstract concept.

9. This is paradoxical. Nietzsche asserts that all representations are accompanied by the basic form of pleasure-and-pain-sensations and yet he develops a hierarchy of representations distinguishing those in the form of pleasure and pain as more basic than those "not" in this form. Apparently representations are a mix of both forms. Those more pervasively constituted by bodily sensation are more fundamental.

10. Cf., Kofman 1993.

11. To renounce metaphysics is not the same as to be free of it. Nietzsche denounces metaphysics but in the same breath, invokes it. Indeed language, as Eugen Fink points out, is itself metaphysical: "What Nietzsche lacks even more decisively (for his work of destroying metaphysics) is a suitable language. What he wants, in fact, is something he cannot yet formulate. For language itself is metaphysical" (Fink 1965, 181, qtd. in Blondel 1991, 201).

12. Here my claim is indebted to a similar assertion by Kofman (1993). She develops this idea in terms of the loss of the "proper" rather than the loss of "identity."

13. "[T]he entire system of the body is called into play, not the mere symbolism of the lips, face, and speech but the whole pantomime of dancing, forcing every member into rhythmic movement. Then the other symbolic powers suddenly press forward, particularly those of music, in rhythmics, dynamics and harmony. To grasp this collective release of all the symbolic powers, man must have already attained that height of self-abnegation which seeks to express itself symbolically through all these powers" (BT 2).

14. "Conceptual" here corresponds to Aristotle's categories (substance, genus, species, relation, quantity, quality, etc.) and subcategories. I prefer "conceptual" to "categorical" because of possible ambiguity with the Kantian connotation of the latter.

15. This inversion indicates *Nietzsche's* view of Aristotle, which reflects a dominant interpretation of Aristotle in the West. By placing "Aristotelian" in quotes, I intend to distinguish this pervasive interpretation from Aristotle's ideas per se.

16. Cf. Martin Heidegger's criticism of the Aristotelian's tendency to view "categories as some sort of encasements into which we stuff beings" (1931/1995, 4).

17. In "On Music and Words" Nietzsche speaks of "*one* primal cause infathomable to us" (1964, 31). See also Sections 4 and 17 in *The Birth of Tragedy*.

18. See the Vorbemerkung by Mazzino Montinari in *Friedrich Nietzsche: Sämtliche Werke, Kritische Studienausgabe in 15 Einzelbänden* (1988, vol.1, 8)

19. Debates continue about whether Nietzsche is neo-Kantian (Cox 1998, 49–51) and if so, what sort. My neo-Kantian reading of Nietzsche agrees with Javier Ibáñez-Noé's in "Nietzsche and Kant's Copernican Revolution: On Nietzsche's Subjectivism" (2002). Ibáñez-Noé's critique of Nietzsche's subject renders "*substantive, causative, singular, and conscious* character of mental life unwarranted" and sees Hume and Kant as "precursors" of Nietzsche, "deconstruction and postmodernism" (2002, 136–137). Christoph Cox argues against a neo-Kantian reading of Nietzsche, which he construes narrowly, with a requisite positing of "the 'real' and the 'apparent' world" (1998, 50)—a view he attributes to Hans Vai-

hinger (1924), Arthur Danto (1965), Julian Young (1992), and Stephen Houlgate (1993). More broadly conceived neo-Kantian readings of Nietzsche connecting Nietzsche's perspectivism and Kant's transcendental subjectivity mentioned by Ibáñez-Noé, are those by Jaspers, Heidegger, Fink, and Kaulbach (2002 132). Cox's view is problematic in supposing an epistemic and absolute rejection on Nietzsche's part of a noumenal world. Nietzsche certainly does, most of the time at least, speak against any noumenal theory (i.e., "The 'apparent' world is the only one: the 'real' world has only been *lyingly added*" (TI "'Reason' in Philosophy" 2), but for aesthetic not epistemic reasons. Cox's reading requires that Nietzsche categorically deny the existence of a noumenal world. Such a pronouncement does not work for Nietzsche; indeed, "What [has Nietzsche] to do with refutations" (GM preface 4)? I do however agree with Cox's conclusions that for Nietzsche, the doctrine of becoming "is intended to counter the metaphysical preoccupation with being, stasis, and eternity by foregrounding the empirically evident ubiquity of change in the natural world, and . . . that becoming describes the incessant shift of perspectives and interpretations in a world that lacks a grounding essence" (Cox, "Nietzsche's Heraclitus and the Doctrine of Becoming," *The Philosopher's Index* 1940–2004), but these need not preclude the variant of neo-Kantianism offered earlier.

20. See Ernst Haeckel's *Die Natur als Künstlerin* (Berlin: Vita Deutsches Verlagshaus, 1924); *Kunstformen der Natur* (Leipzig: Bibliographisches Institute, 1904). For further discussion see Moore 2002, 89–96.

21. Nietzsche designates *übertragen* (to transpose, carry over, transfer, or translate) as the axis of all art and creativity. "Through it the poet transposes his dream" thereby generating myth and cultural health (KSA 7, 395). This operates like Nietzsche's use of *Übertragen* and *Verwandeln* in the following passage. It is "the dramatic artist who transforms (*verwandelt*) himself into alien bodies and talks with their alien tongues and yet can project this transformation (*Verwandlung*) into written verse that exists in the outside world on its own. What verse is for the poet, dialectical thinking is for the philosopher. . . . And just as for the dramatist words and verse are but the stammering of an alien tongue, needed to tell what he has seen and lived, . . . just so every profound philosophic intuition expressed through dialectic and through scientific reflection is the only means for the philosopher to communicate what he has seen. But it is a sad means; basically a metaphoric and entirely unfaithful translation (*Übertragung*) into a totally different sphere and speech" (PTG 3).

22. In his later works, Nietzsche generally rejects the concept of the subject altogether and describes human beings as constitutions of forces continually undergoing bodily and intellectual change.

23. Because *The Will to Power* is a posthumous collection of entries from Nietzsche's notebooks written between 1883–1888, references to the text will include the aphorism number followed by a slash and the year the entry was written.

24. Here are two of many instances of the well-known idea that Nietzsche's writing shows human subjectivity and embodiment eluding understanding: "What do human beings really know about themselves? Are they even capable of perceiving themselves in their entirety just once, stretched out as in an illuminated glass case? Does nature not remain silent about almost everything, even about our bodies, banishing and enclosing us within a proud, illusory consciousness, far away from the twists and turns of the bowels, the rapid flow of the blood stream and the complicated tremblings of the nerve-fibres? Nature has thrown away the key" (TL 142).

"Present experience has, I am afraid, always found us "absent-minded": we cannot give our hearts to it—not even our ears! Rather, as one divinely preoccupied and immersed in himself into whose ear the bell has just boomed with all its strength the twelve beats of noon suddenly starts up and asks himself: 'what really was that which just struck?' so we sometimes rub our ears *afterward* and ask, utterly surprised and disconcerted, 'what really was that which we have just experienced?' and moreover: 'who *are* we really?' and, afterward, as aforesaid, count the twelve trembling bell-strokes of our experience, our life, our *being*—and alas! miscount them.—So we are necessarily strangers to ourselves, we do not comprehend ourselves, we *have* to misunderstand ourselves" (GM preface 1).

25. The following discussion regarding a union and separation existing between the ideal and the body is indebted to Eric Blondel (1991, 204–214).

26. I was directed to this citation by Blondel (1991, 204), who references it *La volonté de puissance*, I, 257, #188. Additional text evidence indicating Nietzsche's rejection of both mechanicism and spiritualism, respectively, as conceptually proper explanations is: "But the finest light beams of nerve activity appear on a surface. . . . It is thus that these pictures are related to the moving nerve-activity beneath. The most tender of all swinging and trembling. . . . All of thought comes to us as a surface and biased, so too our desires" (KSA 7, 446).

27. Blondel mentions Spinoza in this context (1991, 205).

28. For my purposes, I will treat these as synonyms for Nietzsche.

29. This is particularly evident in *The Birth of Tragedy*.

30. Here Moore analyzes Burke's text, *A Philosophical Enquiry into the Origin of our Ideas of the Sublime and the Beautiful*, 2d ed. (London: Dodsley, 1759), part IV, sections vii and xix.

31. Examples of such works are Hippolyte Taine's *Philosophie de l'art* (1865) (a work that, as Moore indicates, Nietzsche owned); Konrad Lange's *Das Wesen der Kunst*, 2 vols. (1901); Herbert Spencer's *The Principles of Psychology* (1872), and

Grant Allen's *Physiological Aesthetics* (1877). For more details see Moore 2002, 85–89.

32. The following inquiry about the significance of the body as a plural unity and as fundamental interpretation is indebted to Blondel 1991, 206–214.

33. "[S]o that, between the original determination and the actual performance of the thing willed, a whole world of new things, conditions, even volitional acts, can be interposed without snapping the long chain of the will. But how much all this presupposes!" (GM 2, 1).

34. By the term "moral subject," I mean a principle of human nature presupposing a causal relation between human desire and human Good. This happens for instance in a common reception of Plato's *Symposium* in which *eros* (desire) seeks Beauty. Beauty is co-extensive with knowledge and the Good. Thus, human desire is linked to a principle of rationality in the sense that if the moral subject properly adapts to the supposed causally necessitated rules of reason, his or her desires will generally be met.

35. These are adaptations of Blondel's "two consequences" resulting from the idea of a concept of body *as* interpretation: "(a) a constitution (for example a categorical constitution) is replaced by an interpretation based on the body and drives; (b) the conscious spirit and intellect becomes the *instrument* of an unconscious interpreting body. Therefore, the former dualism is not simply overturned. There is an *interpretation (a detour)* and *a change of order*" (Blondel 1991, 206).

36. The assertion that the body is fundamental because all human experience appears to begin through the body is not metaphysically based. This is because the concept of body that I attribute to Nietzsche is itself non-foundational for reasons that I develop in this chapter. The body transposes object forces and is itself a transposition of forces of external world chaos. In this sense, Nietzsche's concept of the body qua lived is not in itself "proper." Although Nietzsche may often believe that the unconscious body is the source of art and cultural health, he recognizes this assumption rests on the unstable ground of another: that human experience as and through metaphor seems to be first and foremost deceiving. Thus even his assertion that all begins through the body is subject to question.

37. One will recall Nietzsche's metaphors implying not only that the intellect thinks but that the affects do too.

38. My view here challenges Günter Figal's reading of Nietzsche and those like his that attribute to the world's forces a decisive structure or quality according to which human experience, for Nietzsche, has occurred, does occur and will occur. "The thinking which occurs within" the basic strife of "being and becoming, presence and time" is metaphysics (Figal 1998, x). According to Figal,

Nietzsche's thought indicates a consistent and absolute limit—what he calls "strife"—that makes thought possible. "[S]trife," says Figal, "speak[ing] with Heraclitus, is the father of all things, the necessary coherence of each and every essential thing with its opposite" (1998, x).

39. The implied repetition in a dialectic of saying and unsaying is reminiscent of Freud's concept of reenactment. For an interesting discussion considering Nietzsche's repetition of the reactive mode of evaluation in light of Freud's therapeutic concept of reenactment, see Siemens (2001).

7. NIETZSCHE AFTER NIETZSCHE

1. For the purposes of this chapter "body" and "corporeity" are synonyms and refer to any lived body, that is, to any animal body.

2. By "communication" I mean signification. An example of primordial communication occurs in the cellular activity of all but the simplest animals when cells move (interrogate) and perceive (respond to the movement) (N 179; 219). A cell's perceptive response could be said to signify to surrounding cells. What does it signify? The movement it has just now perceived. An example of complex communication is human language and its various practices—reading, writing, speaking, and thinking—that depend upon an alphabetic system of phonetic writing (a signifying system that signifies sounds of the spoken words signified). Defining communication, for Merleau-Ponty, is traditionally a multifaceted task. Merleau-Ponty commentators, such as James Risser and Duane Davis, have alternately chosen either not to define language, allowing its varying uses to appear contextually in their papers (Risser 1993) or to explicitly designate its varying significations (Davis 1991). Davis, for instance, designates "'language' to mean the physical phenomenon of a system of signifiers, and language (without single quotes) to mean the full event of human experience" (1991, 43).

3. According to Merleau-Ponty, human experience suggests that psychical and physiological functions commingle with one another, almost inextricably along an "intentional arc" (PP 135–136). Terms often used to describe Merleau-Ponty's concept of intentional arc—that is, "reaching for," "directed at," "desire for" (PP 55; 11; 135–136)—imply an open and only partially outlined future toward which is aimed. By "desire" for Merleau-Ponty, I mean such indeterminate aiming.

4. There are three exceptions. An article by Janice Mclane (1996) and two dissertations (Denton 2000; Hetherington 2002).

5. If analyses of Merleau-Ponty's writings with respect to a concept of trauma are rare, those analyzing his texts with regard to topics somewhat related

to trauma, that is, violence (Alcoff 2000; Finn, 1996; O'Neill 1984; Somerville 1971; Whiteside 1991)—or loosely related to trauma, that is, sanity, schizophrenia or psychiatry, are only a fraction less so (Hamrick 1982–1983; Jenner and De Koning 1982; Levin 1982–1983).

6. This is Uexküll's position as discussed by Merleau-Ponty in his second course on nature (N 2003). Examples of simple animals are the marine worm, the medusa (a small jellyfish) and sea urchin. Merleau-Ponty distinguishes simple animals from other more complex animals to which he attributes an *Umwelt* with which animals adapt, respond, and negotiate. Simple animals by contrast, he says, are machine-like because they are virtually closed-off from any *Umwelt*; they are generally incapable of relating, adapting to and negotiating with their environs. To the extent that this is so, suggests Merleau-Ponty, simple animals do not have an *Umwelt*.

7. Robert Vallier (2001, 203; 205).

8. This argument owes much of its inspiration to ideas expressed by Robert Vallier (2001).

9. Cf. Heidegger *Being and Time* 1996, 432, n. 30

10. This thesis and some of the ideas substantiating it, are indebted to Robert Vallier (2001, 198–199; 202).

11. This reference comes from Robert Vallier (2001, 201).

12. This is a variant translation by Robert Vallier (2001, 201) and does not appear in Vallier's 2003 translation of *Nature: Course Notes from the Collège de France*.

13. Vallier (2001, 198).

14. Although Merleau-Ponty hierarchizes orders of animals in *The Structure of Behavior* (1942/1983), by the time of his lecture courses on nature (1956–1960), he suggests a horizontal, co-communicating "empathy" between human bodies and animal bodies (Vallier 2001, 205).

15. Goldstein's major work is *The Organism*, first published in German in 1934 and in English translation in 1939. "Neurologists spend their lives confronting complexities—the bewildering, complex ways in which patients behave and adapt and react. Encountering this process between the world wars, Goldstein tried to unite two traditions, the classical tradition of localizationist neurology and a noetic or holistic approach to the behavior of the organism as a whole. 'The conflict of these two traditions,' Luria wrote in his obituary of Goldstein, 'was the basic content of his life, the attempt to construct a new neurology which had to include the truth of both was his endeavour'" (Sacks 1995, 12–13). Other works by Goldstein pertinent to our discussion include *Zeigen und Greifen* (1931); *L'Analyse de l'aphasie et l'essence du langage* (1933); *Brain Injuries in War* (1942).

16. By "trauma" I mean either physical or psychological trauma, or both. I define psychological trauma as exposure to conditions that overwhelm one's aptitudes for having and processing habits of experience. No single set of conditions is in itself determinately traumatic for all humans and "[n]o two people have identical reactions, even to the same event" (Herman 1997, 58). Even so, "the likelihood that a person will develop a post-traumatic stress disorder depends primarily on the nature of the traumatic event" (Herman 1997, 58). War combat and other situations that place one in an unusually vulnerable psychological and physical position have been shown to affect many people psychologically in predictable ways (Herman 1997, 58). I define physical trauma as exposure to conditions that overwhelm one's anatomical aptitudes to independentally adapt to the conditions (i.e., tissue, spinal cord, brain or organ wounds, serious illness). For evidence of relatively predictable anatomical responses to physical trauma see Barbara Bullock, *Pathology: Adaptation and Alterations in Function* (1996) and Martha Freeman Somers, *Spinal Cord Injury: Functional Rehabilitation* (2001). The idea that persons' responses to trauma happen "across several interpenetrating levels" (i.e., across so-called physical and psychological planes) is mine and inspired by Merleau-Ponty's concept of perception.

17. An exception is a discussion of E. S. Russell's study of animal tissue repair in *Nature: Course Notes from the Collège de France* (178–183).

18. High-level (cervical) spinal injuries cause one to lose sweat gland function because sweat glands are controlled by the sympathetic nervous system whose base of operations—the thoracic and lumbar spinal region—is below the cervical region, and thus can no longer communicate with the brain. In the same way one loses motor and sensory functions of limbs controlled by spinal nerves located below the level of the injury, one loses other functions, like sweat glands, if they are likewise controlled by spinal nerves located below the level of the injury.

19. Patterson carries on the tradition of A. M. DiGiorgio whose studies of 1929, 1943, and 1947 were among the first suggesting spinal-cord memory. Patterson summarizes this tradition in terms of the work of his predecessor: "DiGiorgio had shown that a cerebellar lesion in anesthetized animals could produce an asymmetrical posture in the hind limbs, with one flexed actively and the other extended. If the animal was left in this posture for several hours, then the spinal cord severed at the mid back, the postural asymmetry would remain. The flexion of the limb was assumed to have been "fixated" in the spinal cord by the altered outflows from the cerebellar damage. This spinal fixation was one of the initial demonstrations of memory in the spinal cord" (Patterson 2001, 78).

20. Wolpaw shows via experiments involving the spinal stretch reflex (SSR) and corresponding electrical reflex (H-reflex), both of which are con-

trolled by spinal cord, that the spinal cord "plays an important part in skill acquisition" (2001, 101). His findings suggesting that the spinal cord has memory and learns, challenge traditional views. Wolpaw notes that studies about "the acquisition of motor skills traditionally focus on supraspinal areas such as cerebral cortex and cerebellum" (2001, 101). He compares the reflex acquisition capacity of the spinal cord to the language acquisition aptitude of the cerebellum. "Like the changes in spinal reflexes . . . this plasticity is particularly prominent early in life (e.g., Kuhl, 1998). While languages can be learned, and spinal reflexes can be modified, later in life this learning is clearly constrained by the patterns established in the first few years. A language acquired later on is usually spoken with an accent derived from the individual's original language, and reflex conditioning later in life probably cannot reestablish the antagonist excitation lost early in a normal childhood" (Wolpaw 2001, 119).

21. Wernig, Nanassy, and Müller have shown that the spinal cords of persons chronically confined to wheelchairs due to incomplete para- or quadriplegia, can acquire motor skills for walking (2001, 225–240).

22. By "other" I intend two meanings: the changed circumstances, particularly the new boundaries of environment respective of the new territory of self; and perceivers external to oneself.

23. Elaine Scarry first suggested this reasoning to me in her distinction between the human process of "making up" and "making real" in response to pain, *The Body in Pain: The Making and Unmaking of the World* (1985).

24. In this essay the term "speech" names what I call complex communication as opposed to primordial communication. A precondition of complex communication is a more intricate neural network like that of the human brain. Neural processes located below the cerebrum are comparatively less complex (Grau & Joynes 2001, 46). If for Merleau-Ponty speech and anatomy are both communication—that is, forms of signification—not all communication exists as spoken. "Speech" for my purposes here means language. It accomplishes thought through spoken or written words.

25. For helpful discussion of dissociation, see May Benatar (1995, 318–319).

26. By "linguistic representation," I mean "speech"—that is, the interactive and integrative processes of complex communication (see n. 24 above); not literal or identical representation of an object in either a Cartesian or Kantian sense. Merleau-Ponty's writings have been said to offer a "nonrepresentational" view of perception. Please note that this latter use of "representational" is different than mine. The latter is defined with respect to ontologies that presuppose a subject-object dualism (i.e., those of Descartes, Kant). According to "representation" understood in this second way—representation of a decidedly

"external" world— Merleau-Ponty's writings suggest that our perceptual aptitudes do not represent, but rather, enact interactions with the world. This interactionist reading is implied by James Risser (1993, 134) and aligns with the three-tiered process affording self-signification I describe: 1) interrogating and negotiating an environment; 2) differentiating inner from outer; and 3) presenting one's body surface as self-signifying. Such a tri-fold process could be said to be interactionist because it involves a reciprocal shaping across the planes of so-called "subject" (animal) and "object" (environment) thereby undermining both Cartesian dualist and Kantian rationalist representational theory.

27. A Nietzschean critique of the value of truth would put into question the benefit of "keeping 'true' to the trauma," if one's motivation for doing so were solely the desire for 'truth,' and not primarily a desire for life enhancement.

28. Cf., Cathy Caruth's chapter "Literature and the Enactment of Memory (Duras, Resnais, *Hiroshima mon amour*)" (Caruth 1996).

29. Merleau-Ponty 1994, 248.

30. Cf. Henri Bergson's *Matter and Memory* (1908/1991). See especially Bergson's distinction between "regressive memory" (representational and contemplative) and "progressive memory" (habitual and learned) (1908/1991, 79–83).

31. Here, by "perspectivalism," I mean the idea that so-called "truth" reduces to the perspective of the perceiver.

32. I was first introduced to this language by Sarah Kofman, *Nietzsche and Metaphor* (1993).

33. With respect to Nietzsche's corpus, Eric Blondel speaks about metaphor as an infinite depth of surfaces (1991).

34. I borrow this term from John Thatamanil (2002).

35. By phonetic speech I mean speech and thought among people whose medium for writing involves an alphabet—a set of images symbolizing the sounds of speech: consonant and vowel sounds.

8. NIETZSCHE BEFORE NIETZSCHE

1. David Abram (1996).

2. "Phonetic writing" means writing rooted in notation systems that signify sounds. By putting together combinations of such symbols (letters of an alphabet) one can signify the sounds corresponding to a word that in turn says its referent. Before the development of phonetic writing in the West, humans generally used ideogramic or pictographic writing—writing whose notations appeal to vision rather than sound and whose images represent ideas (ideograms) or things (pictograms).

3. By "literacy" I mean reading and writing phonetic script.

4. Were I to do this, I would compare Nietzsche's Heracliteanism in *Twilight of the Idols* alongside *Philosophy in the Tragic Age of the Greeks*; such an analysis Christoph Cox has already done quite well (1998).

5. Nietzsche resisted the scholarly and scientific specialization of his philological peers, or what had "degenerated to a[n] 'industry of ants'" (*Notizen zu Wir Philologen,* KSA 8:32, NF 3 (63), qtd. in Wischke 2002).

6. This phrase is David Abram's (1996).

7. "'Reason' in Philosophy" (Nietzsche 1889/1968b) will henceforth be abbreviated RP.

8. Common knowledge among scholars of Heraclitus is the obscurity of both the fragments and the man. Little is known biographically about Heraclitus of Ephesus living circa the sixth century B.C.E. (Maly & Sallis 1980, viii). It is even undecided whether the book entitled "On Nature" that Diogenes Laertius says Heraclitus wrote, was ever written. Some suspect that the fragments may have originated orally and were transcribed later (viii). That the fragments come to us imbedded in the disparate contexts of subsequent discourses, with little or no trace of their originating context, continues to pose fundamental challenges and ranges of possibilities for interpreters.

9. Because "logos" has so many significations, that is, speech, conversation, proportion, measure, agreement, a treaty, worth or esteem, the truth of the matter, and so on, the idea here is that if it arises in contexts in Homer as well as in contexts in, for instance, Heraclitus or Plato, which suggest, say, "measure" as opposed to one of the other meanings, that indeed how one interprets "measure" regardless of whether the text is by Homer, Heraclitus, or Plato, would not vary much.

10. An exciting result of factoring in the preliteracy-to-literacy transition is that a traditionally influential use of formal logic for reading Greek texts must be ruled out. A dominating interpretation of the formal logic of the Academy (i.e., of "Platonism" and "Aristotelianism"— placed in quotes to indicate these terms refer to dominant Western receptions of Plato's and Aristotle's writings, and not necessarily Plato or Aristotle's ideas per se) placed the early dictionary compilers in the Roman and Byzantine periods under its spell. To clarify the meanings of verbs and nouns, the Greek lexicon followed the analytical method derived from an emerging logic of fourth-century B.C.E. Greece, formalized in the Scholastic/Aristotelian tradition (Havelock 221–222). This long-standing interpretation of Aristotle's logic still influences the way some practice the Greek lexicon system today. Its methodology, however, has become increasingly challenged. One can no longer assume that the significations of Greek words and grammar remain as constant from the eighth-to-fourth centuries B.C.E. as previously assumed. Indeed, one might consider the four-century span binding

Francis Bacon's English to that of Judith Butler. It can be argued that there is less a need to factor in shifts in English grammar, structures, and practice when examining sixteenth-to-twentieth-century C.E. English literature than eighth-to-fourth century B.C.E. Greek literature. Given the transition to literacy happening in ancient Greece, reading the Greek of both Homer and Plato, respectively, demands a more complex shift in interpretive strategy than reading the English of both Bacon and Butler.

11. C.f. David Abram's chapter "Animism and the Alphabet" (1996).

12. For a detailed discussion, see David Abram's chapter entitled "Animism and the Alphabet" (1996).

13. Cf., Abram 1996, 110.

14. That is to say, we "cannot" given our present perceptual limits and orientation.

15. Charles Scott (1990, 18–35) discusses the middle-voice verb tense of early Sanskrit and Ancient Greek. He describes it as a verb construction that communicates nonreflexively, that is, without a clear distinction between a subject and object. Scott describes Nietzsche's concept of self-overcoming as the "middle-voice of metaphysics" by comparing its metaphysics to nonreflexive sentence structures (Scott 1990, 18).

16. A transitive verb is a verb that takes an object.

17. I identify Heraclitus fragments using Diels-Kranz numeration; each Diels-Kranz fragment number is preceded with a "D." (Diels & Kranz 1974).

18. In order to reference different translations of a single fragment, at the end of the English translation I usually signify the translator parenthetically—either by full name or by an abbreviation—followed by the translator's number for the fragment. T. M. Robinson uses the Diels-Kranz numeration, and so, there is no number following his abbreviated name. The translator abbreviations are: "K" for Charles Kahn (1979); "K/R" for G. S. Kirk and J. E. Raven (1983); "R" for T. M. Robinson (1987); and "W" for Philip Wheelwright (1964).

19. My mode of interpretation approaches that of Martin Heidegger and Eugin Fink in one important respect. In the Heraclitus Seminar they speculate about "what is unsaid in what is said" in some Heraclitus fragments (Heidegger & Fink 1970/1993, viii). For those accustomed to much English-language commentary on Heraclitus, Heidegger's and Fink's aggressively speculative method might seem ponderous or even reckless. Like Heidegger and Fink, nevertheless, I aim not at "an interpretation limited to a recapitulation" of what Heraclitus explicitly said, for such "can never be a real explication" (Heidegger 1929/1962, 206, qtd. in Heidegger & Fink 1970/1993, viii). Rather, "what is essential in all philosophical discourse is not found in the specific propositions of which it is

composed but in that which, although unstated as such, is made evident through these propositions" (Heidegger 1929/1962, 206, qtd. in Heidegger & Fink 1970/1993, viii).

20. Unless otherwise indicated, translations for fragments below are by Kevin Robb (1986). For a discussion defending these fragments as the "genuine psyche fragments" see Robb (1986, 325–326).

21. For this particular category, I am indebted to Kevin Robb (1986, 325).

22. This translation of the fragment from Plutarch's "Concerning the Face which appears in the Orb of the Moon," in *Moralia* 28 is by Harold Cherniss (1958, 943 D–E). John Sallis adopts this as a literal translation in "Hades, Heraclitus Fragment B 98" (1980, 61).

23. Robb lists two translations, one similar to Cherniss's and this more satirical translation which invites Martha Nussbaum's reading: "Dumb *anthropoi* still believe in a Homeric *psyche* which, if in Hades is blind and must find its way around by sniffing. Ridiculous!" (1972, 156).

24. My discussion of the genealogy of *psyche* from the eighth to the fourth centuries B.C.E. is rooted in Kevin Robb's analysis (1986, 318). Due to the limits of my project, I only sketch his *psyche* genealogy.

25. Cf., Robb 1986, 315–351.

26. Robb (1986, 343) lists other scholars who agree that the late fifth century B.C.E. marks the earliest date for the concept of a unitary personal soul. "An early work stressing the importance of this period was John Burnet's 'The Socratic Doctrine of the Soul,' *Proceedings of the British Academy* VII (1916). Support was added in 1933 (first American edition 1953) from his Edinburgh colleague, A. E. Taylor, in *Socrates: The Man and His Thought* (Garden City, NY: Doubleday, 1953), pp. 134ff" (Robb 1986, 343). Robb lists additional scholars supporting this view: "Jan Bremmer, *The Early Greek Concept of the Soul* (Princeton, NJ: Princeton University Press, 1983); E. A. Havelock, *Preface to Plato* (Cambridge, MA: Harvard University Press, 1963; reprinted 1983" (1986, 343).

27. A concept of *psyche* can be called "unified," if it has intelligent desire, and "disunified," if it does not. Put differently, a concept of a unified *psyche* alleges that humans have the potential to *know* what they ultimately want *and* that that which is wanted is consistent and rational (i.e., is knowledge of the Good), and not by contrast, logically at odds with itself. Judith Butler describes it this way, "If desire were a principle of irrationality, then an integrated philosophical life would be chimerical, for desire would always oppose this life, undermine its unity" (Butler 1987, 3).

28. Compare Waugh 1991, 616.

29. The exact date of Heraclitus's birth remains unknown, but he "was in his prime about 500 B.C.E." (Freeman 1948, 24).

30. I set "Plato's" in scare quotes to indicate that this term refers to a dominant Western reception of Plato's writings.

31. Derrida discusses a relation between *pharmakon* and its multiple significations. These are blocked by "an effect of 'Platonism,' the consequence of something already at work in the translated text, in the relation between 'Plato' and his 'language'" (1972/1981, 98).

32. According to Derrida, Plato's *Phaedrus* privileges the meaning "remedy" (1972/1981).

33. Waugh 1991, 610; Abram 1996, 93–136, 225–274.

34. An important difference in the syntax of an oral tradition, contends Havelock, is its attunement to human psychological conditions that enhance the memory's ability to remember a story. Such epic "syntax," writes Havelock, "repeats and reports all information so far as possible in the form of concrete and particular events which happen in sequence, not as propositions which depend on each other in logical connexion" (1982, 226). Verse characterized by particular concrete occurrences provides an instrument with which a people can preserve a culture in the living memories of human beings (Havelock 1982, 226–227). Because Homeric-to-Platonic Greek registers not only the beginning of a literate tradition but the transpiration of an oral one, its texts also record this speaking shift.

35. "The Problem of Socrates" (Nietzsche 1889/1968b) will henceforth be abbreviated PS.

References

Abbey, Ruth. 1999. *The Roots of* Ressentiment: Nietzsche on Vanity. *New Nietzsche Studies* 3 (3/4): 47–61.

Abram, David. 1996. The Spell of the Sensuous: Perception and Language in a *More-Than-Human World*. New York: Vintage Books.

Alcoff, Linda Martin. 2000. Merleau-Ponty and Feminist Theory on Experience." In *Chiasms: Merleau-Ponty's Notion of Flesh*, edited by Fred Evans.

Allen, Grant. 1877. *Physiological Aesthetics*. London: Henry S. King.

American Psychiatric Association. 1994. *Diagnostic and Statistical Manual of Mental Disorders: DSM IV*. Washington, D.C.: American Psychiatric Association.

Andrew, Edward. 1999. The Cost of Nietzschean Values. *New Nietzsche Studies* 3 (3/4): 63–76.

Ansell-Pearson, Keith. 1997. *Viroid Life: Perspectives on Nietzsche and the Transhuman Condition*. London: Routledge.

Aristotle. 1941. *Poetics*. Translated by Ingram Bywater. In *The Basic Works of Aristotle*, edited by Richard McKeon. New York: Random House.

——. 1979. *Aristotle's Metaphysics*. Translated by Hippocrates G. Apostle. Grinnell, Iowa: Peripatetic Press.

Babich, Babette E. 1990a. On Nietzsche's Concinninity. *Nietzsche-Studien* 19: 71–73.

——. 1990b. Nietzsche and the Condition of Postmodern Thought: A Post-Nietzschean Postmodernism. In *Nietzsche as Postmodernist*, edited by Clayton Koelb. Albany: State University of New York Press.

Behler, Ernst. 1995. Nietzsche's Study of Greek Rhetoric. *Research in Phenomenology* XXV: 3–26.

Being John Malkovich. 2000. VHS (ca. 113 min.): sd., col.; 1/2in. Calif.: USA Home Entertainment.

Benator, May. 1995. Running Away from Sexual Abuse: Denial Revisited. *Families in Society* 76 (5): 315–320.

Bergson, Henri. 1991. *Matter and Memory*. Translated by N. M. Paul & W. S. Palmer. New York: Zone Books. Originally published in 1908.

Blondel, Eric. 1991. *Nietzsche: The Body and Culture*. Translated by Seán Hand. Stanford, Calif: Stanford University Press. Originally published in 1986.

Buckley, Thomas, & Alma Gottlieb, eds. 1988. *Blood Magic: The Anthropology of Menstruation*. Berkeley: University of California Press.

Bullock, Barbara L. 1996. *Pathophysiology: Adaptation and Alterations in Function*. 4th ed. Philadelphia: Lippencott-Raven Publishers.

Burke, Edmund. 1759. *A Philosophical Enquiry into the Origin of our Ideas of the Sublime and the Beautiful*, 2nd ed. London: Dodsley

Butler, Judith. 1987. *Subjects of Desire*. New York: Columbia University Press.

Capps, Lisa, & Elinor Ochs. 1995. *Constructing Panic: The Discourse of Agoraphobia*. Cambridge, Mass.: Harvard University Press.

Carpenter, Rhys. 1933. The Antiquity of the Greek Alphabet. *American Journal of Archaeology* 37: 8–29.

——. 1938. The Greek Alphabet Again. *American Journal of Archaeology* 42: 58–69.

Caruth, Cathy. 1996. *Unclaimed Experience: Trauma, Narrative, and History*. Baltimore, Md.: Johns Hopkins University Press.

Cox, Christoph. 1998. Nietzsche's Heraclitus and the Doctrine of Becoming. *International Studies in Philosophy* 30: 3, 49–63.

Danto, Arthur. 1965. *Nietzsche as Philosopher*. New York: Columbia University Press.

Darwin, Charles. 1985. *Origin of the Species*. Harmondsworth, UK: Penguin. Originally published in 1859.

Denton, Wendy. 2000. The Soul Has Bandaged Moments: Self-Injury as a Language of Pain and Desire. Ph.D. diss., Graduate Theological Union, Berkeley, Calif.

Derrida, Jacques. 1974. *Of Grammatology*. Translated by Gayatri Chakravorty Spivak. Baltimore, Md.: John Hopkins University Press. Originally published in 1967.

——. 1981. Plato's Pharmacy. In *Dissemination*, translated by Barbara Johnson. Chicago: University of Chicago Press. Originally published in 1972.

Diels, Hermann, & Walther Kranz. 1974. *Die Fragmente der Vorsokratiker*, vol. 1. Zürich: Weidmann.

Donnellan, Brendan. *Nietzsche and the French Moralists*. Bonn: Bouvier.

Figal, Günter. 1998. *For a Philosophy of Freedom and Strife*. Translated by Wayne Klein. Albany: State University of New York Press. Originally published in 1994.

Figert, Anne E. 1996. *Women and the Ownership of PMS: The Structuring of a Psychiatric Disorder*. New York: Aldine de Gruyter.

Fingarette, Herbert. 1972. *Confucius: The Secular as the Sacred*. New York: Harper & Row.

Fink, Eugin. 1965. *Le philosophie de Nietzsche*. Paris: Les Éditions.

Finn, Geraldine. 1996. *Why Althusser Killed His Wife: Essays on Discourse and Violence*. Atlantic Highlands, N.J.: Humanities Press.

Fonrobert, Charlotte Elisheva. 2000. *Menstrual Purity: Rabbinic and Christian Reconstructions of Biblical Gender*. Stanford, Calif.: Stanford University Press.

Foot, Philippa. 1991. Nietzsche's Immoralism. *New York Review of Books*. 38(11): 18–22.

Foucault, Michel. 1977. Nietzsche, Genealogy, Morality. In *Language, Counter-Memory, Practice*. Ithaca, N.Y.: Cornell University Press. Originally published in 1971 in *Hommage à Jean Hyppolite*.

——. 1979. *Discipline and Punish*. New York: Random House. Originally published in 1975.

——. 1980. *The History of Sexuality*. Vol. 1, *An Introduction*. New York: Random House. Originally published in 1976.

——. 1988. *The History of Sexuality*. Vol. 3, *The Care of Self*. New York: Random House. Originally published in 1984.

——. 1990. *The History of Sexuality*. Vol. 2, *The Use of Pleasure*. New York: Random House. Originally published in 1984.

Freeman, Kathleen. 1948. *Ancilla to the Pre-Socratic Philosophers*. Cambridge, Mass.: Harvard University Press.

Gatens, Moira. 1996. *Imaginary Bodies: Ethics, Power and Corporeality*. New York: Routledge.

Geertz, Clifford. 1983. *Local Knowledge: Further Essays in Interpretive Anthropology*. New York: Basic Books.

Gillan, Garth. 1988. Foucault and Nietzsche: Affectivity and the Will to Power. In *Postmodernism and Continental Philosophy*, edited by Hugh Silverman and Don Welton, Albany: State University of New York Press.

Glenn, Paul. 2001. The Great Health: Spiritual Disease and the Task of the Higher Man. *Philosophy and Social Criticism* 27 (2):100–117.

Goldstein, Kurt. 1933. "L'Analyse de l'aphasie et l'essence du langage." *Journal de psychologie normale et pathologique* 30: 430–451.

——. 1942. *Brain Injuries in War*. New York: Grune & Stratton.

——. 1995. *The Organism*. New York: Zone Books. Originally published in 1934.

Grau, James, & Robin Joynes. 2001. Pavolovian and Instrumental Conditioning within the Spinal Cord: Methodological Issues. In *Spinal Cord*

Plasticity: Alterations in Reflex Function, edited by Michael M. Patterson & James W. Grau. Boston: Kluwer Academic Publishers.

Greenberg, Moshe. 1995. The Etymology of נדה "(Menstrual) Impurity." In *Solving Riddles and Untying Knots: Biblical, Epigraphic and Semitic Studies in Honor of Jonas Greenfield*, edited by Z. Zevit, S. Gitin, & M. Sokoloff. Winona Lake, Ind.: Eisenbrauns.

Grosz, Elizabeth. 1996. *Space, Time and Perversion*. New York: Routledge.

Guthrie, W. K. C. 1962. *A History of Greek Philosophy: The Earlier Presocratics and the Pythagoreans*. Cambridge: Cambridge University Press.

Haar, Michel. 1979. Nietzsche and Metaphysical Language. In *The New Nietzsche*, edited by David B. Allison. New York: Delta.

Haeckel, Ernst. 1924. *Die Natur als Künstlerin*. Berlin: Vita Deutsches.

——. 1904. *Kunstformen der Natur*. Leipzig: Bibliographisches Institute.

Hamrick, William S. 1982–1983. Language and Abnormal Behavior: Merleau-Ponty, Hart, and Laing. *Review of Existential Psychology and Psychiatry* 18, 181–203.

Hansen, Hardy, & Gerald Quinn. 1987. *Greek: An Intensive Course*. New York: Fordham University Press.

Harries, Karsten. 1988. The Philosopher at Sea. In *Nietzsche's New Seas*, edited by Michael Allen Gillespie & Tracy B. Strong. Chicago: University of Chicago Press.

Havelock, Eric. 1982. Preliteracy and the Presocratics. In *The Literate Revolution in Greece and Its Cultural Consequences*. Princeton, N.J.: Princeton University Press.

——. 1983. The Linguistic Task of the Presocratics. In *Language and Thought in Early Greek Philosophy*, edited by Kevin Robb. La Salle, Ill.: Hegeler Institute and the Monist Library of Philosophy.

——. 1986. *The Muse Learns to Write: Reflections on Orality and Literacy from Antiquity to the Present*. New Haven, Conn.: Yale University Press.

Hegel, G. W. F. 1979. *Phenomenology of Spirit*. Translated by A. V. Miller & J. N. Findlay. Oxford: Oxford University Press. Originally published in 1807.

Heidegger, Martin & Eugin Fink. 1993. *Heraclitus Seminar*. Translated by Charles S. Seibert. Evanston, Ill.: Northwestern University Press. Originally published in 1970.

Heidegger, Martin, 1961. *Nietzsche*, Vol. 1. Tübingen, Germany: Verlag Günther Neske Pfullingen.

——. 1962. *Kant and the Problem of Metaphysics*. Translated by James S. Churchill. Bloomington: Indiana University Press. Originally published in 1929.

——. 1972. Aletheia (Heraclitus, Fragment B16). In *Early Greek Thinking*, translated by David Farrell Krell & Frank A. Capuzzi. San Francisco: Harper & Row. Originally published in 1954 in *Vorträge und Aufsätze*.

——.1995. *Aristotle's* Metaphysics Θ *1–3: On the Essence of Actuality of Force*. Translated by Walter Brogan & Peter Warnak. Bloomington: Indiana University Press. Originally published in 1931.

——. 1996. *Being and Time*. Translated by Joan Stambaugh. Albany: State University of New York Press. Originally published in 1927.

Herman, Judith. 1993,1997. *Trauma and Recovery: The Aftermath of Violence–From Domestic Abuse to Political Terror*. New York: Basic Books.

Hetherington, Dan. 2002. Disaster Trauma: A Phenomenological-Linguistic Analysis of Buffalo Creek Flood Accounts. Ph.D. diss., Duquesne University, Pittsburgh, Pa.

Hisamatsu, Hoseki Shin'ichi. 1984. *Die Fülle des Nichts*. Translated by Takashi Hirata & Johanna Fischer. Pfullingen, Germany: Neske Verlag.

Horowitz, Gregg M. 2001. *Sustaining Loss: Art and Mournful Life*. Stanford, Calif.: Stanford University Press.

Houlgate, Stephen. 1993. Kant, Nietzsche and the "Thing in itself." *Nietzsche-Studien* 22: 133, 135.

Hughes, Fiona. 2002. Nietzsche's Janus Perceptions and the Construction of Values. *Journal of the British Society of Phenomenology* 33(2): 116–137.

Janaway, Christopher. 1997. Nietzsche's Illustration of the Art of Exegesis. *European Journal of Philosophy* 5(3): 251–268.

Jenner, F. A., and A. J. J. De Koning (eds.). 1982. *Phenomenology and Psychiatry*. London: Grune & Stratton.

Kahn, Charles H. 1979. *The Art and Thought of Heraclitus: An Edition of the Fragments with Translation and Commentary*. Cambridge: Cambridge University Press.

Kirk, G. S., J. E. Raven, & M. Schofield. 1983. *The Prescocratic Philosophers*. 2nd ed. Cambridge: Cambridge University Press.

Kofman, Sarah. 1988. Baubô: Theological Perversion and Fetishism. In *Nietzsche's New Seas*, edited by Michael Allen Gillespie & Tracy B. Strong. Chicago: University of Chicago Press.

——. 1993. *Nietzsche and Metaphor*. Translated by D. Large. Stanford, Calif.: Stanford University Press. Originally published in 1983.

Laclau, Ernesto. 1996. Deconstruction, Pragmatism, Hegemony. In *Deconstruction and Pragmatism*, edited by Chantal Mouffe. London: Routledge.

Lacoue-Labarthe, Phillippe. 1971. Le detour (Nietzsche et la rhétorique). *Poétique* 5.

Lange, Konrad. 1901. *Das Wesen der Kunst*, 2 vols. Berlin: Grote.

Laws, Sophie. 1990. *Issues of Blood*. London: Macmillan.

Laycock, Stephen W. 1994. *Mind as Mirror and the Mirroring of Mind: Buddhist Reflections on Western Phenomenology*. Albany: State University of New York Press.

Leigh, James A. 1978. Deleuze, Nietzsche and the Eternal Return. *Philosophy Today*, Fall: 213–216.

Leiter, Brian. 2000. Nietzsche's Metaethics: Against the Privilege Readings. *European Journal of Philosophy*, (December) 8: 3 .

Levin, David Michael. 1982–1983. Sanity and Myth in Affective Space: A Discussion of Merleau-Ponty. *Philosophical Forum (Boston)*. 14 (Winter): 157–189.

Lifton, Robert Jay. 1993. *The Protean Self: Human Resilience in an Age of Fragmentation*. New York: Harper & Row.

Lingis, Alphonso. 1979. The Will to Power. In *The New Nietzsche*, edited by David B. Allison. New York: Delta.

Longino, Helen. 1990. *Science as Social Knowledge: Values and Objectivity in Scientific Inquiry*. Princeton, N.J.: Princeton University Press.

Magnus, Bernd. 1986. Nietzsche's Philosophy in 1888: *The Will to Power* and the Übermensch. *Journal of the History of Philosophy* 24(1): 79–98.

Magnus, Bernd, & Kathleen M. Higgins, eds. 1996. *Cambridge Companion to Nietzsche*. Cambridge: Cambridge University Press.

Magnus, Bernd, Stanley Stewart, & Jean-Pierre Mileur. 1993. *Nietzsche's Case: Philosophy as/and Literature*. New York: Routledge.

Maly, Kenneth, & John Sallis, eds. 1980. *Heraclitean Fragments: A Companion Study to the Heraclitus Seminar*. Tuscaloosa: University of Alabama Press.

McLane, Janice. 1996. "The Voice on the Skin: Self-Mutilation and Merleau-Ponty's Theory of Language." *Hypatia* 11(4) (Spring): 107–118.

Medina, Joyce. 1995. *Cézanne and Modernism: The Poetics of Painting*. Albany: State University of New York Press.

Merleau-Ponty, M. 1962. *Phenomenology of Perception*. Translated by Colin Smith. London: Routledge & Kegan Paul. Originally published in 1945.

——. 1970 & 1988. "Themes from the Lectures." Translated by J. O'Neill. In *In Praise of Philosophy and Other Essays*. Evanston, Ill.: Northwestern University Press. Originally published in 1968.

——. 1983. *The Structure of Behavior*. Translated by Alden L. Fisher. Pittsburgh, Pa.: Duquesne University Press. Originally published in 1942.

——. 1994. *La Nature: Notes Cours Du Collège de France*. Paris: Seuil.

——. 2003. *Nature: Course Notes from the Collège de France*. Compiled and with notes by Dominique Séglard. Translated by Robert Vallier. Evanston, Ill.: Northwestern University Press. Originally published in 1994.

Montinari, Mazzino. 1988. Vorbemerkung. In *Friedrich Nietzsche: Sämtliche Werke, Kritische Studienausgabe in 15 Einzelbänden*, Vol. 1, edited by Giorgio Colli and Mazzino Montinari. Berlin: Deutscher Taschenbuch Verlag/Walter de Gruyter.

Moore, Gregory. 2002. *Nietzsche, Biology and Metaphor*. Cambridge: Cambridge University Press.

Müller-Lauter, Wolfgang. 1974. Welt als Wille zur Macht. *Tijdschrift voor Philosophie*, Vol. 36: 78–106.

Murphy, Tim. 2001. *Nietzsche, Metaphor, Religion*. Albany: State University of New York Press.

Nehamas, Alexander. 1985. *Nietzsche: Life as Literature*. Cambridge, Mass.: Harvard University Press.

Nietzsche, Friedrich. 1894–1904. *Nietzsche's Werke*, 15 vols. Leipzig: Naumann.

——. 1947. *La volonté de puissance*. Translated by G. Bianquis, 32nd ed., 2 vols. Paris: Gallimard. Originally published in 1906.

——. 1962–1969. *Philosophy in the Tragic Age of the Greeks*. Translated by Marianne Cowan. Chicago: Henry Regnery. Originally published in 1894.

——. 1964. On Music and Words. In *The Complete Works of Friedrich Nietzsche*, translated by Oscar Levy. New York: Russell & Russell. Written in 1871 and published posthumously.

——. 1966. *Beyond Good and Evil*. Translated by Walter Kaufmann. New York: Random House. Originally published in 1886.

——. 1967. *The Birth of Tragedy*. Translated by Walter Kaufmann. New York: Random House. Originally published in 1872.

——. 1968a. *The Will to Power*. Translated by Walter Kaufmann & R. J. Hollingdale. New York: Random House. Originally published in 1906.

——. 1968b. *Twilight of the Idols*. In *Twilight of the Idols/The Antichrist*, translated by R. J. Hollingdale. New York: Penguin Books. Originally published in 1889.

——. 1968c. *The Anti-Christ*. In *Twilight of the Idols/The Anti-Christ*, translated by R. J. Hollingdale. New York: Penguin Books. Originally published in 1895.

——. 1974. *The Gay Science*. Translated by Walter Kaufmann. New York: Random House. Originally published in 1882.

——. 1978. *Thus Spoke Zarathustra*. Translated by Walter Kaufmann. New York: Penguin Books. Originally published in 1883.

——. 1982. *Daybreak: Thoughts on the Prejudices of Morality*. Translated by Walter Kaufmann. Cambridge: Cambridge University Press. Originally published in 1881.

——.1983. *On the Uses and Disadvantages of History for Life*. In *Untimely Meditations*, translated by R. J. Hollingdale. Cambridge: Cambridge University Press. Originally published in 1874.

——. 1988. *Friedrich Nietzsche: Sämtliche Werke, Kritische Studienausgabe in 15 Einzelbänden.* Edited by Giorgio Colli & Mazzino Montinari. Berlin: Deutscher Taschenbuch Verlag/Walter de Gruyter.

——. 1989. *On the Genealogy of Morals.* Translated by Walter Kaufmann & R. J. Hollingdale. New York: Random House. Originally published in 1887.

——. 1994. *On the Genealogy of Morality.* Translated by Carol Diethe. Cambridge: Cambridge University Press. Originally published in 1887.

——. 1999. On Truth and Lying in a Non-Moral Sense. In *The Birth of Tragedy and Other Writings.* Translated by Ronald Speirs and edited by Raymond Geuss & Ronald Speirs. Cambridge: Cambridge University Press. Written in 1872 and published posthumously.

Nishitani, Keiji. 1982. *Religion and Nothingness.* Translated by Jan Van Bragt. Berkeley: University of California Press.

Noé, Ibáñez. 2002. Nietzsche and Kant's Copernican Revolution. *New Nietzsche Studies* 5:1/2 (Spring/Summer): 132–149.

Nussbaum, Martha. 1972. *Psyche* in Heraclitus I. *Phronesis* 17.

Oliver, Kelly. 1995. *Womanizing Nietzsche: Philosophy's Relation to the Feminine.* New York: Routledge.

Olson, Carl. 2000. *Zen and the Art of Postmodern Philosophy.* Albany: State University of New York Press.

Ondaatje, Michael. 1992. *The English Patient.* New York: Vintage Books.

O'Neill, John. 1984. Merleau-Ponty's Critique of Marxist Scientism. In *Phenomenology and Marxism,* edited by Bernhard Waldenfels. London: Routledge.

Parry, Milman. 1928. *L Èpithète Traditionalle dans Homèr.* Paris: Société d' éditions "Les Belles Lettres."

——. 1971. Whole Formulaic Verses in Greek and Southslavic Heroic Song. In *The Making of Homeric Verse: The Collected Papers of Milman Parry,* edited by Adam Parry. Oxford: Clarendon Press.

Patterson, Michael M. 2001. Spinal Fixation: Long-term Alterations in Spinal Reflex Excitability with Altered or Sustained Sensory Inputs. In *Spinal Cord Plasticity: Alterations in Reflex Function,* edited by Michael M. Patterson & James W. Grau. Boston: Kluwer Academic Publishers.

Paz, Octavio. 1994. Hygiene and Repression. In *Ourselves Among Others,* edited by Carol J. Verburg. Boston: Bedford Books.

Plato. 1925. *Lysis, Symposium, Gorgias.* Translated by W. R. M. Lamb. Cambridge, Mass.: Harvard University Press.

——. 1937. *The Republic: Books I–V.* Translated by Paul Shorey. Cambridge, Mass.: Harvard University Press.

Plutarch. 1958. "Concerning the Face which Appears in the Orb of the Moon." In *Moralia* 28, Loeb, Vol. 12, translated by Harold Cherniss. Cambridge, Mass.: Harvard University Press.

Ray, Sumana. 1990. *Vegetarisches aus Indien*. Hamburg: Merit-Verlag.

Rey, Jean-Michel. 1988. Commentary. In *Nietzsche's New Seas*, edited by Michael Allen Gillespie & Tracy B. Strong. Chicago: University of Chicago Press.

Ripper, Margie. 1991. A Comparison of the Effect of the Menstrual Cycle and the Social Week on Mood, Sexual Interest, and Self-Assessed Performance. In *Menstruation, Health and Illness*, edited by Diana L. Taylor & Nancy F. Wood. New York: Hemisphere.

Robb, Kevin. 1986. Psyche and Logos in the Fragments of Heraclitus: The Origins of the Concept of Soul. *The Monist*. 69 (July): 315–351.

Robinson, Andrew. 1995. *The Story of Writing*. New York: Thames & Hudson.

Robinson, T. M. 1987. *Heraclitus Fragments: A Text and Translation with a Commentary*. Toronto: University of Toronto Press.

Rodin, Mari. 1992. The Social Construction of Premenstrual Syndrome. *Social Science and Medicine*. 35 (1): 49–56.

Rosenblatt, Roger. 1996. New Hopes, New Dreams. *Time*, 26 August.

Rousseau, Jean-Jacques. 1979. *Emile: or On Education*. Translated by Allan Bloom. New York: Basic Books. Originally published in 1762.

Ruden, Sarah. 1998. Staring at Yellow and Green (Life and Art and PMS). *Michigan Quarterly Review*. 37 (3): 409–425.

Russell, E. S. 1946. *Directiveness of Organic Activities*. Cambridge: Cambridge University Press.

Sacks, Oliver. 1995. Foreword. In *The Organism*, by Kurt Goldstein. New York: Zone Books.

Sallis, John. 1980. Hades: Heraclitus, Fragment B 98. In *Heraclitean Fragments: A Companion Study to the Heraclitus Seminar*, edited by Kenneth Maly & John Sallis. Tuscaloosa: University of Alabama Press.

——. 1986. *Being and Logos: The Way of the Platonic Dialogue*. Atlantic Highlands, N.J.: Humanities Press.

——. 1991. Dionysus: Resounding Excess. In *Crossings*. Chicago: University of Chicago Press.

Saussure, Ferdinand de. 1959. *Course in General Linguistics*. Translated by Wade Baskin. New York: McGraw-Hill. Originally published in 1916.

Scarry, Elaine. 1985. *The Body in Pain: The Making and Unmaking of the World*. Oxford: Oxford University Press.

Schacht, Richard. 1983. *Nietzsche*. London: Routledge & Kegan Paul.

Schiller, Friedrich. 1994. *On the Aesthetic Education of Man*. Translated by Reginald Snell. Bristol, UK: Thoemmes. Originally published in 1867.

Schopenhauer, Arthur. 1969. *The World as Will and Representation*, Vol. 1. Translated by E. F. J. Payne. New York: Dover. Originally published in 1844.

Schrift, Alan. 1988. Genealogy and/as Deconstruction. In *Postmodernism and Continental Philosophy*, edited by Hugh Silverman & Dan Welton. Albany: State University of New York Press, 1988.

——. 1989. Nietzsche and the Critique of Oppositional Thinking. *History of European Ideas*. 11: 783–790

Scott, Charles. 1990. *The Question of Ethics: Nietzsche, Foucault, Heidegger*. Bloomington: Indiana University Press.

Seigfried, Charlene Haddock. 1975. Why are Some Interpretations Better than Others. *New Scholasticism*. 1975 (Spring): 140–161.

Shapiro, Gary. 1990. Translating, Repeating, Naming: Foucault, Derrida, and *The Genealogy of Morals*. In *Nietzsche as Postmodernist*, edited by Clayton Koelb. Albany: State University of New York Press.

Siemens, Herman W. 2001. Nietzsche's Agon with Ressentiment: Towards a Therapeutic Reading of Critical Transvaluation. *Continental Philosophy Review*. 34: 69–93.

Simmer-Brown, Judith. 2001a. Dharma Talk. Boulder Vipassana Community, July 2001, Boulder Colorado.

——. 2001b. *Dakini's Warm Breath: The Feminine Principle inTibetan Buddhism*. Boston: Shambhala.

Sjöholm, Cecilia. 2001. The Expression of Another in Me. *Chiasmi International* 3: 173–185.

Somers, Martha Freeman. 2001. *Spinal Cord Injury: Functional Rehabilitation*. 2nd ed. Upper Saddle River, N.J.: Prentice Hall.

Somerville, John. 1971. Violence, Politics and Morality. *Philosophy and Phenomenological Research*. 71 (32): 241–249.

Sophocles. 1982. *Sophocles: Three Tragedies*. Translated by H. D. F. Kitto. New York: Oxford University Press.

Spencer, Herbert. 1872. *The Principles of Psychology*, 2 vols. London: Williams & Norgate.

——. 1879. *The Data of Ethics*. Lodnon: Williams & Norgate.

Stambaugh, Joan. 1999. *The Formless Self*. Albany: State University of New York Press.

Steinem, Gloria. 1983. If Men Could Menstruate. In *Outrageous Acts and Everyday Rebellions*. New York: Holt, Rinehart & Winston.

Stotland, Nada L., & Harwood, Bryna. 1994. Social, Political and Legal Considerations. In *Premenstrual Dysphorias: Myths and Realities*, edited by Judith H. Gold & Sally K. Severino. Washington, D.C.: American Psychiatric Press.

Sullivan, Shannon. 2001. *Living Across and Through Skins: Transactional Bodies, Pragmatism and Feminism*. Bloomington: Indiana University Press.

Taine, Hippolyte. 1865. *Philosophie de l'art*. Paris: Baillière.

Takeda, Sumio. 2000–2001. *New Nietzsche Studies* 4(3/4): 99–105.

Thatamanil, John T. 2002. Beyond Number: On the Relational Possibilities of Tillich's Symbolic and Speculative Trinitarianism. Paper presented at the Ninth International Paul Tillich Symposium, 1 June, Frankfurt, Germany.

Todorov, Tzvetan. 1984. *The Conquest of America*. Translated by Richard Howard. New York: Harper & Row. Originally published in 1982.

Vaihinger, Hans. 1924. Nietzsche and His Doctrine of Conscious Illusion. *The Philosophy "As If."* Translated by C. K. Ogden. New York: Harcourt, Brace. Originally published in 1911.

Vallier, Robert. 2001. The Indiscernible Joining: Structure, Signification, and Animality in Merleau-Ponty's *La nature. Chiasmi International* 3: 187–212.

Van der Kolk, Bessel A., & Onno Van der Hart. 1995. "The Intrusive Past: The Flexibility of Memory and the Engraving of Trauma." In *Trauma: Explorations in Memory*, edited by Cathy Caruth. Baltimore, Md.: Johns Hopkins University Press.

Vlastos, Gregory. 1955. On Heraclitus. *American Journal of Philology*, 76: 357. Reprinted in G. Vlastos. 1995. *Studies in Greek Philosophy*, Vol. 1, Princeton, N.J.: Princeton Univeristy Press.

Waugh, Joanne. 1991. Heraclitus: The Postmodern Philosopher? *The Monist*. 74 (October): 603–623.

Wernig, Anton, Andras Nanassy, & Sabina Müller. 2001. Laufband (Treadmill) Therapy in Incomplete Para- and Tetraplegia. In *Spinal Cord Plasticity: Alterations in Reflex Function*, edited by Michael M. Patterson & James W. Grau. Boston: Kluwer Academic Publishers.

Wheelwright, Philip. (1964). *Heraclitus*. New York: Atheneum.

White, Dick. 1988. Heidegger on Nietzsche: The Question of Value. In *Postmodernism and Continental Philosophy*, edited by Hugh Silverman & Dan Welton. Albany: State University of New York Press.

Whiteside, Kerry, H. 1991. Universality and Violence: Merleau-Ponty, Malraux, and the Moral Logic of Liberalism. *Philosophy Today*. 35 (4) (Winter): 372–389.

Wischke, Mirko. 2002. Nietzsche and Neo-Kantianism: On Gadamer and Philology as an Untimely Reflection. *New Nietzsche Studies*. 5(1/2) (Spring/Summer): 97–112.

Wittgenstein, Ludwig. 1961. *Tractatus Logico-Philosophicus*. Translated by D. F. Pears & B. F. McGuinness. London: Routledge & Kegan Paul.

Wolpaw, Jonathan. 2001. Spinal Cord Plasticity in the Acquisition of a Simple Motor Skill. In *Spinal Cord Plasticity: Alterations in Reflex Function*, edited by Michael M. Patterson & James W. Grau. Boston: Kluwer Academic Publishers.

Young, Julian. 1992. *Nietzsche's Philosophy of Art*. Cambridge: Cambridge University Press.

Zita, Jacqueline N. 1989. The Premenstrual Syndrome 'Dis-easing' the Female Cycle. In *Feminism and Science*, edited by Nancy Tuana. Bloomington: Indiana University Press.

Index